Dynamic and Seamless Integration of Production, Logistics and Traffic

Eberhard Abele • Manfred Boltze •
Hans-Christian Pfohl
Editors

Dynamic and Seamless Integration of Production, Logistics and Traffic

Fundamentals of Interdisciplinary Decision Support

Springer

Editors
Eberhard Abele
Institute of Production Management, Technology and Machine Tools
Technische Universität Darmstadt
Darmstadt, Germany

Manfred Boltze
Chair of Transport Planning and Traffic Engineering
Technische Universität Darmstadt
Darmstadt, Germany

Hans-Christian Pfohl
Supply Chain and Networkmanagement
Technische Universität Darmstadt
Darmstadt, Germany

ISBN 978-3-319-41095-1 ISBN 978-3-319-41097-5 (eBook)
DOI 10.1007/978-3-319-41097-5

Library of Congress Control Number: 2016954240

Printed on acid-free paper

This Springer imprint is published by Springer Nature
The registered company is Springer International Publishing AG Switzerland

Contents

Interdisciplinary Research in Production, Logistics and Traffic: Introduction to Dynamo PLV and Overview of the Book

Eberhard Abele, Manfred Boltze, and Hans-Christian Pfohl

Introduction to Dynamo PLV

A size of almost 3 km^2, 40,000 people, over 1000 trucks a day and a regularly frequented train connection. What sounds like a small city with heavy inbound and outbound traffic is the description of a modern automobile plant which produces more than 2000 vehicles a day and matches the characteristics of various OEM production facilities all over the globe. Several thousands of suppliers, a big portion of them from overseas, are supplying this type of production facilities while there are tens of thousands of different parts. Given the high volatility and especially the high complexity of the automobile production—when ordering a German premium car one can choose from up to one million variations of seat configurations alone—those parts cannot be supplied in a push process. Instead, they have to be pulled through a well-orchestrated supply chain guaranteeing the availability of the right part in the right quality at the right time.

This example from the automotive industry highlights the current and future challenges of value-added networks. Trends like globalisation, increasing volatility, digitalisation and urban growth have an increasing impact on daily life and

E. Abele
Institute of Production Management, Technology and Machine Tools, Technische Universität Darmstadt, Darmstadt, Germany
e-mail: abele@ptw.tu-darmstadt.de

M. Boltze
Chair of Transport Planning and Traffic Engineering, Technische Universität Darmstadt, Otto-Berndt-Straße 2, 64287 Darmstadt, Germany
e-mail: boltze@verkehr.tu-darmstadt.de

H.-C. Pfohl (✉)
Supply Chain and Networkmanagement, Technische Universität Darmstadt, Darmstadt, Germany
e-mail: pfohl@scnm.tu-darmstadt.de

E. Abele et al. (eds.), *Dynamic and Seamless Integration of Production, Logistics and Traffic*, DOI 10.1007/978-3-319-41097-5_1

business. This can be observed through higher congestion of transport infrastructure as well as the increasing importance of logistics services or 24/7 manufacturing. In order to cope with those trends, companies have to move from optimising their own operation to an integration of supply chain activities (Christopher and Peck 2012). This implies that instead of companies, in future, supply chains or even geographical regions will compete against each other.

To ensure competitiveness, companies and public authorities have to increasingly coordinate their decisions. But what is often referred to as the extended enterprise in the literature (Bititci et al. 2005), meaning a close collaboration of a company and its suppliers with the goal to increase the returns for each supply chain partner, is only seldom implemented in practice. Instead, in order to react to existing challenges, decision-makers from companies or public authorities implement decision support methods and instruments as well as various measures that allow them to optimise a limited subsystem (Abele et al. 2012; Pfohl et al. 2012; Zuber et al. 2015). For example, banning trucks from inner city streets may solve social conflicts (inside the city as a limited subsystem) but may result in longer transportation routes generating economical and ecological disadvantages (in the overall system). Integrated solutions on the supply chain level are often inhibited due to conflicting goals of supply chain partners and their lack of willingness to share critical information. Those conflicts often have their sources at the level of interdisciplinary conflicts of competing departments like logistics, supply chain management and purchasing (Wang et al. 2015). While challenges on the supply chain level already seem hard to solve, the decisions of public authorities do have to be added to the equation in order to achieve a true understanding of the indentation of production, logistics and traffic and to form true interdisciplinary solutions.

To address those challenges, the project Dynamo PLV was founded in 2011 and funded by the Hessian Ministry of Science and Arts (HMWK) through the LOEWE (Landes-Offensive zur Entwicklung Wissenschaftlich-ökonomischer Exzellenz) programme from 2011 until 2014. Dynamo PLV is an acronym, standing for Dynamic and Seamless Integration of Production, Logistics and Traffic. The goal of the project was to identify and systemise the interdisciplinary relations between production, logistics and traffic and to develop concepts for integrated decision support for companies and public authorities. Since the project team's overall target was sustainable research on the integration of production, logistics and traffic at TU Darmstadt and EBS, the following vision was verbalised:

Dynamo PLV builds the foundation for an international research and transfer platform to promote the holistic, interdisciplinary, sustainable as well as flexible integration of production, logistics and traffic by developing and implementing state-of-the-art technology.

The project team was formed by an interdisciplinary research team from the Technische Universität Darmstadt and the European Business School (EBS). Participating researchers from the TU Darmstadt were Prof. Eberhard Abele (project spokesman), Prof. Hans-Christian Pfohl (project coordinator), Prof. Manfred Boltze, Prof. Ralf Elbert, Prof. Hanno Friedrich, Prof. Joachim Metternich, Prof. Herbert Meyr, Prof. Andreas Oetting and Prof. Alejandro Buchmann. Participating

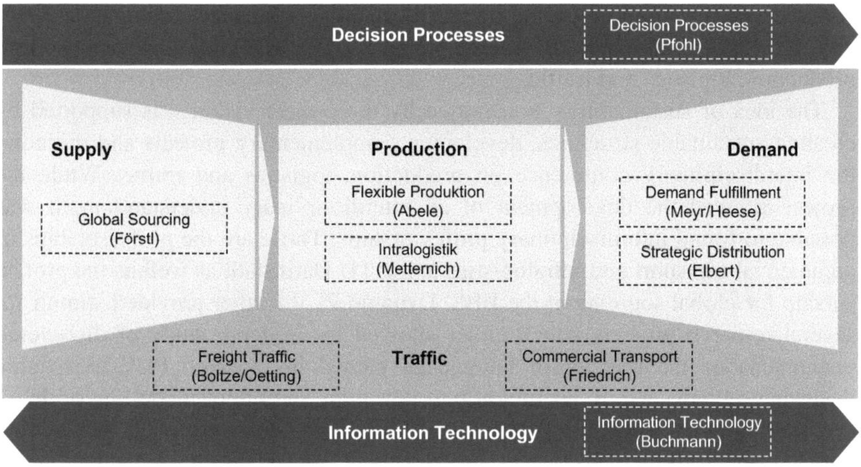

Fig. 1 Project structure—Dynamo PLV

researchers from the EBS were Prof. Kai Förstl and Prof. Heese. In addition, more than 20 Ph.D. students from both Universities were involved in Dynamo PLV.

The overall project structure of Dynamo PLV is depicted in Fig. 1. In order to tackle the challenge of integrating production, logistics and traffic in an interdisciplinary research project, TU Darmstadt and EBS had to build adequate structures and shape an interdisciplinary project terminology. Research on interdisciplinary artefacts requires that "interdisciplinary researchers leave their comfort zone and go beyond the boundaries of their own fields" (Sanders and Wagner 2011, p. 321). Therefore, interdisciplinary working groups were formed in order to address specific issues. The underlying objectives when setting up those working groups were:

- Problem isolation: Shaping an interdisciplinary understanding and selecting central interdisciplinary decision processes for further investigation.
- Systematisation: Structuration of decision variables in the various disciplines, identification of qualitative interrelations and analysis of the role of underlying goals from flexibility and sustainability perspectives.
- Operationalisation: Identification of quantitative interrelations between the subsystems of production, logistics and traffic and development of concepts for quantitative model integration.

The interdisciplinary working groups build the frame for the cooperation of the sub-projects illustrated in Fig. 1. Interdisciplinary results provided by the working groups were the interdisciplinary decision navigator and the interdisciplinary decision map as presented in chapter "The Interdisciplinary Decision Map: A Reference Model for Production, Logistics and Traffic". Together, they represent an instrument for mapping interdisciplinary decisions and studying interrelations. The interdisciplinary decision navigator also represents a platform hosting the overall project results. The working groups further provided insights into the effects of goal conflicts and information sharing on supply chain performance (Wang et al. 2015) and developed prototypes of simulation models for integrating decisions from production, logistics and traffic.

The idea of sustainability transported by the project vision was supported by creating sustainable structures, developing complementary projects and initiating the interdisciplinary conference on production, logistics and traffic. While the project allowed the development of an interdisciplinary understanding, it was possible to found interdisciplinary professorships. Those are the professorships of commercial transport and intralogistics at the TU Darmstadt as well as the professorship for global sourcing at the EBS. Dynamo PLV further provided stimuli for several research projects which either allowed for in-depth study of discovered phenomena or the transfer of knowledge gained in Dynamo PLV to specific decision problems, e.g. in air freight transportation, shop floor modelling or delivery time window scheduling. The interdisciplinary conference on production, logistics and traffic (ICPLT) was successfully launched by the project group in 2013. The particular achievement of the conference was to provide an interdisciplinary plenum for researchers as well as practitioners from companies and public authorities. In order to strengthen the German interdisciplinary research landscape, the second edition of the conference was organised in July 2015 in close cooperation with TU Dortmund in Dortmund. The third edition will be organised once again at TU Darmstadt in 2017.

The LOEWE focus Dynamo PLV integrated a research community that provides an essential contribution towards the resolution of complex challenges of our economy. The project provided a solid foundation to fulfil the vision of an interdisciplinary research and transfer platform. The structures shaped at TU Darmstadt and EBS, follow-up projects focusing on interdisciplinary integration, and the continuation of the ICPLT will facilitate an interdisciplinary understanding in production, logistics and traffic. Reaching this type of understanding and integration will be a strong supporting factor for value-added networks to persist in the increasing competition.

Overview of the Book

Besides this introduction chapter, the book starts with two chapters in Part I, which contribute mainly to the fundamentals of interdisciplinary research in production, logistics and traffic.

Part II, in contrast, illuminates in seven chapters the interdisciplinarity from a domain-specific perspective of production, logistics and traffic, which shows the high practical relevance of the topic.

As each chapter represents the results of one working group, a brief summary of each result is given in the following.

Chapter "Interdisciplinary Decisions in Production, Logistics and Traffic and Transport"

Decision-making in value-added networks is becoming increasingly complex. To ensure that decisions made do not counterproductively interfere with each other, interdisciplinary decision-making should be facilitated. While this sounds simple from a theoretical position, practice shows that decisions are usually made in functional silos, ignoring existing interdependencies. So in order to facilitate interdisciplinary decision-making, the chapter firstly draws from descriptive decision theory in order to provide a general structure of barriers of interdisciplinary decision-making. In the second step, a typology of interdisciplinary decision interrelations is derived, which then forms the foundation for a normative decision model. This model offers advice concerning the contingent utilisation of tools for interdisciplinary decision support.

Chapter "The Interdisciplinary Decision Map"

Due to strong interdependencies between production, logistics and traffic, a decision in one of these fields has an impact on the others. However, decision-makers in and around today's supply chains rarely consider effects of their decisions on other participants of the supply chain or the traffic system. Thus, a tool for decision support which clearly illustrates the variety of impacts of a decision is highly desirable. Accordingly, this chapter presents a reference model in the context of production, logistics and traffic, called Interdisciplinary Decision Map (IDM). The IDM allows for describing and analysing interdisciplinary impacts of decisions across the disciplines. Thus, it can serve as decision support tool for decision-makers out of the considered domains. The IDM's applicability is demonstrated by using it to analyse selected impacts of an HGV toll's introduction on production, logistics and traffic.

Chapter "Flexible Design of Lean Production Systems in Response to Fluctuations Due to Logistics and Traffic"

A value-creating supply chain of physical products involves several actors in the fields of production, logistics and traffic. The diversity of these stakeholders often leads to unplanned disruptions in the chain. Industrial production plants, especially those set up according to the principles of lean production, suffer severely from disturbances. One approach for at least partially absorbing the effects is raising the overall flexibility levels within production itself. The focus of this chapter is thus set on how flexibility can be leveraged in lean production systems to deal with disturbances resulting from fluctuations in logistics and traffic.

The chapter starts with the identification of flexibility requirements using the fault-tree analysis method. Following this, a closer look at the term "flexibility" itself in manufacturing contexts as well as on the different types of flexibility will be taken. Volume and mix flexibility are identified as the types most relevant for industrial companies. Thus, different models for measuring volume and mix flexibility in the literature will be reviewed. The chapter then turns to the actions for increasing flexibility in production systems and presents selected technical, organisational and human approaches. The focus is from then on set on lean production systems and their special conditions and prerequisites. First, selected approaches for addressing volume and mix flexibility in lean assembly systems will be discussed. Afterwards, the innovative concept of Cellular Manufacturing as an approach to transfer the ideas from lean assembly to lean machining will be introduced and volume and mix flexibility advantages due to the Cellular Manufacturing concept will be presented. The chapter concludes with a cost comparison of Cellular Manufacturing versus complete machining.

Chapter "Simulation-Based Assessment of Lean Production Methods"

In this chapter, a simulation-based approach will be presented to measure the effects of implementing chosen strategies for increasing volume and variant flexibility of production systems. The chapter covers the strategies of flexible manufacturing systems, production levelling and demand-managed milk run, which are investigated according to their effects on the mentioned flexibility indicators as well as the overall production performance. Three main factors of the production system are addressed in particular: equipment, employees and internal logistics. For these studies, an extensive simulation model of a real production environment was developed, allowing inexpensive validations of further methods and problem-solving approaches.

Chapter "Revenue Sharing Between Production, Logistics and Traffic"

The ongoing trend to outsource logistics services increases the importance of logistic service provider (LSP) as the intermediary between production and traffic. Distribution logistics with the supply chain has a key role in fulfilling customers' demand in time and good quality. To perform their services efficiently and effectively, the LSPs need relationship-specific information. Since information is valuable the customer avoids sharing those with the LSP. To incentive information sharing revenue-sharing contracts can be an instrument to align the interests of the partners. This chapter has the aim to research the influence of fixed revenue sharing schemes on the behaviour of the contract partners regarding the willingness to share information. The applied method is controlled laboratory experiments with students. The results of the experiment show a significant change in the participant's behaviour and indicate to a higher willingness to share information under revenue-sharing contracts.

Chapter "Supply Chain Risk Management in the Automotive Industry"

This chapter endeavours to capture some of the rich dialogues and insights that the research team has generated based on 14 in-depth case studies and literature mining about supply chain risk management in the automotive industry. Three conceptual models adopting cross-functional and multi-tier perspectives are introduced to help companies not only avoid supply chain risks but also seize potential opportunities.

Chapter "Temporal Shifts in Freight Traffic"

In the coming years, the problems of increasing capacity constraints on German roads will increase due to the expected growth of road freight transport. Temporal shift of freight transport may be suitable for a more efficient use of existing infrastructure. This chapter analyses the theoretical potential for a temporal shift of truck traffic by taking the example of selected highway sections in the German Region Frankfurt RheinMain. The analysis shows that peak loads in truck traffic are considerably weaker than in passenger car traffic. One reason is a higher efficiency in truck traffic compared to passenger car traffic. Furthermore, the observed proportion of trucks during peak loads of total traffic is usually below 12 %. The expected contribution of temporal shift of truck trips to avoid congestion is rather low; therefore, possibilities for temporal shift of traffic in peak loads must be primarily searched in passenger car traffic. Opportunities for temporal shifts of truck traffic may result from changes of production and logistics processes, which enable increased transport by night.

Chapter "Freight Transport Demand Management"

In the future, influencing the freight transport demand will become important in light of the increasing freight-transport-related problems. But freight transport is explicitly not the focus of mobility management as the established concept for demand management in the context of traffic management. Thus, until now, traffic management lacks a concept for influencing the freight transport demand, being equal to mobility management. Accordingly, this chapter presents a concept of freight transport demand management within traffic management. The principles for that concept have been derived from mobility management. Furthermore, measures are described which are suitable for influencing the freight transport demand. Finally, further need for research is highlighted.

Chapter "Implications for Freight Transport Demand Modelling from Interdisciplinary Research"

The freight transport system is a major determinant for the competitiveness of logistics and production activities. However, logistics and production shape freight transport demand. Therefore, freight transport demand models are needed that can capture the influence and requirements of ongoing trends in production and logistics. In this chapter, the German automotive industry is used as an example how certain trends might influence the development of freight transport demand. Here, the impact of changes in the supply chain structure on freight transport demand is emphasised. Furthermore, data sources available on the sectoral level are discussed. Based on the insights on ongoing trends, available data and sectoral characteristics, a concept for a sectoral freight transport model for the German automotive industry is developed.

Summary and Acknowledgements The project group has formulated the following vision for the LOEWE project:

Dynamo PLV generates an international research and transfer platform for the promotion of holistic, interdisciplinary, sustainable and flexible connection of production, logistics and traffic under the development and implementation of newest technology.

Thanks to the LOEWE funding and the professorships established through it, as well as the interdisciplinary network, the foundation for the implementation of the vision of a transfer platform for interdisciplinary research in the context of production, logistics and traffic was laid. This applies to the scientific community on the one hand, due to the high visibility—also in the international field—achieved by the continuity of the conference ICPLT. The fact that the conference will be organised in cooperation with the TU Dortmund from now on also contributes to better networking. Visibility and a professional public were not only achieved by the project group through the organisation of the ICLPT. It also published far more than 100 contributions, mainly in the international field, which were presented at international conferences. Furthermore, the visibility in the practical field was strengthened through the involvement of practice partners, especially regarding the economic advisory board. Manufacturing companies, service providers and the public authorities were sensitised to the necessity of interdisciplinary networking and were shown potentials on the basis of the approaches developed by the project group.

First of all, we would like to thank the Hessian Ministry if Science and Art (HMWK), which supported our project within the LOEWE programme. Without such financial support, the initiation of interdisciplinary cooperation would not have been possible. We would also like to thank all those involved in the project for their great commitment to interdisciplinary cooperation. Our special thanks go to Dr. Christian Zuber, head of the business office, for his efficient project management. Even in difficult phases of project processing, he was able to motivate all project team members not to lose sight of the vision. Last but not least, we would like to thank Springer Publishing for giving us the opportunity for this interdisciplinary publication. Compared to purely disciplinary research, it is still much more difficult for interdisciplinary research to find high-ranking publication mediums for the publication of research results. We hope the resulting volume finds a wide readership!

References

Abele, E., Elbert, R., Bachmann, J., Berbner, U., & Özsucu, Ö. (2012). Überwindung von Zielkonflikten in Netzwerken der Automobilindustrie/Wettbewerbsfähigkeit durch Integration von Produktion, Logistik und Verkehr. *Industrie Management, GITO Verlag, Berlin, 28*(5), 29–32.

Bititci, U. S., Mendibil, K., Martinez, V., & Albores, P. (2005). Measuring and managing performance in extended enterprises. *International Journal of Operations and Production Management, 25*(4), 333–353.

Christopher, M., & Peck, H. (2012). *Marketing logistics*. London: Routledge.

Pfohl, H.-C., Berbner, U., & Ehrenhoefer, M. (2012). Fundamental problematic issues in interdisciplinary decision-making: An integrated view on production, logistics, and traffic and transport. In *Proceedings of 9th International Meeting on Logistics Research*, 15–17 Aug 2012, Montreal.

Sanders, N. R., & Wagner, S. M. (2011). Multidisciplinary and multimethod research for addressing contemporary supply chain challenges. *Journal of Business Logistics, 32*, 317–323.

Wang, L., Pfohl, H.-C., Berbner, U., & Keck, A. K. (2015). Supply chain collaboration or conflict? Information sharing and supply chain performance in the automotive industry. In U. Clausen & H. Friedrich (Eds.), *Commercial transport—Essential link for production, logistics and society*. Berlin: Springer.

Zuber, C., Pfohl, H.-C., & Berbner, U. (2015). Integrating domains in supply chains: Development of requirements for interdisciplinary decision support based on an integrative framework of production, logistics, and traffic. In E. D. S. Institute, J. Stentoft, & A. Paulraj (Eds.), *Research in the decision sciences for innovations in global supply chain networks: Best papers from the 2014 annual conference* (pp. 233–254). Financial Times Prentice Hall. Retrieved from http://tubiblio.ulb.tu-darmstadt.de/73611/

Part I
Fundamentals of Interdisciplinary Research in Production, Logistics and Traffic

Interdisciplinary Decisions in Production, Logistics, and Traffic and Transport: Measures for Overcoming Barriers in Interdisciplinary Decision-Making

Hans-Christian Pfohl, Ulrich Berbner, and Christian Zuber

Introduction

Megatrends like decreasing product life cycles, demographic change, and market volatility as well as the progression of digitalization lead to increased complexity in value-added networks (Straube and Pfohl 2008; Blecker and Kersten 2006; Craighead et al. 2007). This increased complexity leads to an ever-growing number of restrictions, goals, and alternatives that have to be observed when making decisions (Vidaillet 2012), but also to an increasing functional specialization of decision makers, which are lacking oversight concerning overall problems (Sanders et al. 2013). In addition, methods and instruments, which should facilitate decision-making in complex situations, usually fail to address an interdisciplinary scope (Boltze et al. 2015). In practice, this often results in decisions being made without taking care or even knowing about potentially triggered follow-up decisions, putting the competitiveness of today's value-added networks at high risk.

While research in supply chain management does prove the positive impact of decision-making abilities on supply chain performance (Gligor et al. 2013), decision theory does usually not build the basis for research in the area of cross-functional integration in supply chains (Frankel et al. 2008; Lambert et al. 2005).

Decisions can be supported by various types of decision support solutions, which are usually developed addressing predefined real-life decision problems. Due to the high complexity and to keep the development of decision support solutions manageable, practical decision problems in supply chain management are simplified by abstraction through eliminating irrelevant variables and interrelations (Blackhurst

H.-C. Pfohl (✉) • U. Berbner • C. Zuber
Supply Chain and Networkmanagement, Technische Universität Darmstadt, Darmstadt, Germany
e-mail: pfohl@scnm.tu-darmstadt.de; berbner@scnm.tu-darmstadt.de; zuber@scnm.tu-darmstadt.de

E. Abele et al. (eds.), *Dynamic and Seamless Integration of Production, Logistics and Traffic*, DOI 10.1007/978-3-319-41097-5_2

et al. 2005). Research also shows that especially in cases where rather complex decisions have to be made, decision makers shall utilize rules of thumb or heuristics instead of complex decision support solutions which are often too time-consuming in their application and often even don't provide good solutions due to their high level of abstraction (Rahman and De Feis 2009).

The development of decision support solutions, and the definition of underlying decision problems in research, is closely tied to the various research disciplines such as information management or operations research (Bohanec 2003). Often, it can be seen as an incremental approach optimizing existing solutions, which usually address stereotypes of theoretical defined problems and application scenarios (Mezias and Starbuck 2008).

The objectives of this chapter are to understand the barriers of interdisciplinary decision-making and to show how various measures in the area of human, organization, and technology can help to overcome interdisciplinary boundaries.

In order to do so, we conducted a case study in the research project Dynamo PLV among logistics service providers, producing companies and traffic authorities in order to collect relevant decision variables from their fields of operation and to sample cases where decision makers do not implement integrated decision-making due to interdisciplinary barriers. Based on the case study, we were able to derive four different types of interdisciplinary barriers. In the second step, we used the empirical data from the cases combined with a literature review in order identify measures to overcome the different barriers. In the third step, we focused on decision support solutions constructing a typology of interrelations as well as a typology of decision support solutions. The combination of these typologies allows matching adequate decision support solutions with practical interdisciplinary problems.

Decision Theory and Interdisciplinary Decision-Making

Interdisciplinary decision-making is already the subject of various research activities. However, research focuses predominantly on general aspects, e.g., the underlying process of interdisciplinary decision-making (Harrison 1993; Koutsoukis and Mitra 2003), or is related to specific application fields like the health sector (Elwyn and Edwards 2009; Strausa et al. 2011). An investigation of (barriers in) interdisciplinary decision-making along the value network regarding the interactions between production, logistics, and traffic and transport did not take place so far.

Decision theory is usually separated into descriptive, prescriptive, and normative decision theory (Dillon 1998; Bell et al. 1988; Laux 2012). While descriptive decision theory is dedicated to describe how decisions are made in practice, prescriptive approaches give answers toward how a decision has to be made. Prescriptive models are usually based on a theoretical foundation, while respecting insights from descriptive decision theory (Dillon 1998). In this chapter, we integrate descriptive and prescriptive concepts by first describing the barriers of

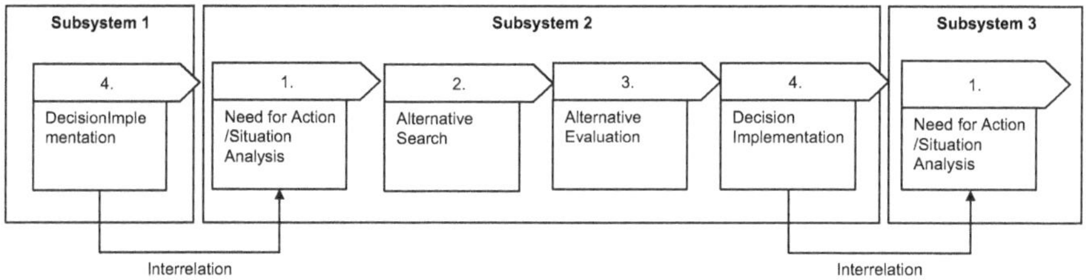

Fig. 1 Decision process and interrelations

interdisciplinary decision-making and second identifying which measures and solutions can be implemented in order to overcome those barriers.

Decision-making is often structured into a four-step process as illustrated in Fig. 1. Those steps comprehend the *identification of need for action*, *alternative search*, *alternative evaluation*, and *implementation* (Simon 1997; Huber 1980). Decisions do usually not stand for themselves inside an organization or value-added network but are highly interrelated (Vidaillet 2012). A decision is usually triggered if the as-is state deviates from a certain target state (Pfohl 1977). In the interdisciplinary context, this kind of imbalance can either be induced by a change of the environment of the decision maker's system (e.g., change of weather prohibiting the utilization of a certain transport mode) or by a decision inside the decision maker's system or inside an adjacent system (e.g., the introduction of road tolls). As shown in Fig. 1, interrelations between decisions can be described by the last step of an initial decision-making process (making decision/implementation) and the first step of a second, interrelated decision-making process (need for action/ situation analysis/intelligence).

While decisions can be supported by measures in the area of organization, people, and technology (Pfohl et al. 2012a, b; Leavitt 1965), this chapter focuses on technologic decision support solutions. Various research disciplines are dedicated to develop decision support solutions (Bohanec 2003). Due to the complexity of decisions made in practice, prescriptive models usually abstract from real-life problems (Blackhurst et al. 2005; Harrison 1993) often resulting in models, which either have a certain distance to practice or are strongly dedicated to support a tight frame of predefined purely disciplinary decisions (Mezias and Starbuck 2008). Various methods and frameworks have been designed in order to match decision problems and decision support solutions. For example, Rahman and De Feis (2009) distinguish decision problems based on the underlying time pressure and the problem's complexity and further allocate optimal measures for decision support.

In order to answer our main research question *how barriers in interdisciplinary decision-making can be resolved*, we have to build up an understanding of interdisciplinary decision-making and interdisciplinary interrelations in production, logistics, and traffic. Therefore, we define three objectives:

1. Objective: Identify types of barriers in interdisciplinary decision-making.
2. Objective: Identify measures to overcome barriers in interdisciplinary decision-making.

3. Objective: Deriving requirements for decision support solutions in an interdisciplinary context and developing a decision support typology that allows to match various decision support solutions to various interdisciplinary contexts.

Research Methodology and Empirical Data

As variables for describing the systems state of production, logistics, and traffic, we use the existing organizational structure, the company's employees, the technology employed, and the tasks carried out which represent the four basic configurable variables of management (Leavitt 1965, 1978). The characteristics of those configurable variables are determined by systems' specific *decisions* and the accrued *tasks*. That leads to organizational centralized or fragmented subunits' production, logistics, and traffic, which are influenced by other subsystems and their environment. The structural couplings between the subsystems determine if and to what degree decision makers and other subsystems can influence a subsystem.

To investigate interdisciplinary interdependencies and the barriers for integrated decision-making, a multi-case study (Yin 2009) was conducted among different companies.

According to Schramm (1971), "the essence of a case study (...) is that it tries to illuminate a decision or set of decisions: why they were taken, how they were implemented, and with what result." Hence, the case study method seems to be adequate in order to describe interconnections between the different disciplines, the underlying decisions and barriers. Yin (2009) describes four different types of case studies, among which one type follows the so-called embedded multi-case design. This type of design was implemented in this study in order to achieve a higher level of robustness than with a single-case design. The case study was prepared, conducted, and evaluated by following these eight steps: (1) definition of the research question, (2) case selection, (3) preparation, (4) field entrance, (5) data analysis, (6) shaping hypotheses, (7) literature comparison, and (8) reaching closure (Eisenhardt 1989).

We derived the following research question based on the insights we gained from literature review:

> Why do decision makers in production, logistics, and traffic and transport often not include/consider affected parties of their value chain in their decision-making processes and how can decision makers be enabled to make more integrated decisions?

Answers to this question enabled us to build up a structured theory about relevant barriers for interdisciplinary decision-making.

We have conducted three main cases (automotive, spare parts/engineering, regional traffic management) and gathered input from about 55 experts (companies, science, public authorities). The entities selected for conducting the case study represented different market segments and industries (B2B/B2C, logistics/production). The case study's objective was mainly to study decision-making processes in the supply chain; hence diverse experts from the areas of production and logistics from companies representing different supply chain tiers were selected and are complemented by experts from the public sector. We especially addressed decision makers who could also give insight about existing decisions' impact on traffic and transport. Cases/entities were added sequentially during the research process in order to replicate but also to extend our emergent theories in the area of interdisciplinary decision-making. Saturation was reached after investigating existing processes in 11 different entities.

Overcoming Barriers in Interdisciplinary Decision-Making

Identification of Barriers in Interdisciplinary Decision-Making

Our investigations confirm that in practice decisions are often made disciplinary, despite increased requirements for interdisciplinary decision-making. Enterprises simplify complex decision situations, by not including certain aspects (like external logistics or traffic influences within production decisions), allowing to focus on a subsystem only. Effects on other subsystems, or the overall system, are usually not considered (functional thinking) or left to the partners within the value network.

It became evident that in some cases, decision outcomes concerning one entity were not transparent to the decision-making entity. This can also mean that the objectives of the affected entities are not transparent to the decider. This situation is illustrated by example EB1 in Table 1, which could be observed in two different cases. Here, the supplier decides on a type of packaging to package all its goods in order to standardize the packaging process and to save packaging costs. The supplier *does not know* that the receiver of the goods (the OEM) needs to repack the goods in the incoming goods department manually since the packaging is not compatible with its automated storage system. If the supplier was informed, he would probably be interested in changing his packaging processes, probably coming along with higher fees which could easily be absorbed by the receiver due to his decreased handling efforts.

It also became apparent that in some cases, the decision-making entity is just *not interested* in considering objectives of other parties involved. An example for this case is example EB2, where the production department decided to use a dedicated type of container for ergonomic reasons. In this case, the deciders were aware of the difficulties this may cause for the other parties involved, like the transport operator or the logistics department, but had the power to enforce their preferred solution.

Table 1 Exemplary field observations

Example		Actual decision maker (marked bold) and decision makers influenced by the decision outcome
EB1	Goods have to be repacked before storage since packaging is incompatible with storage system	• Logistics department supplier (reduction of cost of packing material) • Logistics department OEM (reduction of process complexity)
EB2	Specific types of reusable containers are used for ergonomic reasons even so this causes low transport capacity utilization	• Production OEM (optimization of container ergonomics) • Logistics OEM (capacity utilization) • Transport operator (capacity utilization)
EB3	Truck drivers/transport operators do not interchange information about when a delivery reaches its destination which may cause congestion at the destination site due to ramp overload	• Logistics department OEM (optimum ramp utilization) • Truck driver/transport operator (reduction of loading/unloading times)
EB4	Delivery trucks are blocked by passenger traffic	• Logistics department OEM (optimum road usage) • Different public entities (political objectives)

Of course there were cases observed, where the decision makers were interested in finding interdisciplinary solutions, which might be beneficial for all parties involved. In example EB3, the parties involved would like to reduce congestion caused by the mismanagement of delivery trucks at the ramp. But since there is *no information system available* that holds the required information and provides access for all relevant decision makers, it is not possible to make system-wide or interdisciplinary decisions.

Example EB4 shows that even if the relevant information is available, decisions might still be made in silos. The sample illustrates a case were all parties involved are interested in finding an optimized solution and have all the information necessary to do so, but are not provided with any tools or method to do so. In the sample, delivery trucks are regularly blocked by passenger traffic. This happens at regular hours each day due to shift changeovers at the nearby production site. In order to resolve the problem, different stakeholders from industry and public entities would need to find an integrated solution.

Based on those findings, a systematic overview of barriers for interdisciplinary decision-making is presented in Fig. 2 while the following core propositions can be verbalized:

P1: In interdisciplinary contexts, decisions focus on subsystems since not all outcomes and effects of decisions are transparent to the decision maker.

P2: In interdisciplinary contexts, decisions focus on subsystems since the decision maker is not interested in the impact his decisions might have on other systems.

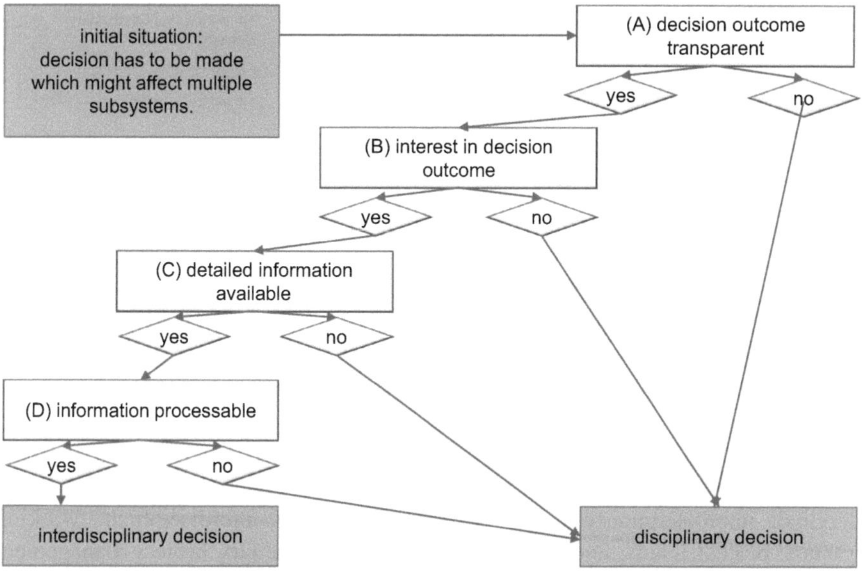

Fig. 2 Barriers for interdisciplinary decision-making

P3: In interdisciplinary contexts, decisions focus on subsystems since not all the relevant information/data is available to the decision maker in a sufficient depth of detail.

P4: In interdisciplinary contexts, decisions focus on subsystems since the available information/data is not processable.

The reasons for the barriers shown in Fig. 2 can be broken down in a number of sub-reasons. For example, non-transparent decision outcomes might result from missing interdisciplinary education or the nonexistence of process documentation, while a lack of interest in system-wide decision outcomes might result from differing objectives of the decision-making entities or from linguistic barriers among interdisciplinary groups. As shown in Fig. 3 and lent by the approach of Leavitt (cf. Leavitt 1965, 1978), the identified sub-reasons can be structured into three interconnected operationally systems: *organization, people*, and *technology*, which influence the decision-making process.

Measures to Facilitate Interdisciplinary Decision-Making

The areas organization, people, and technology represent a starting point for facilitating integrated decision-making and allowed us to structure measures identified in the case study [for more details on the various measures to overcome interdisciplinary barriers, please refer to Pfohl et al. (2012a, b)].

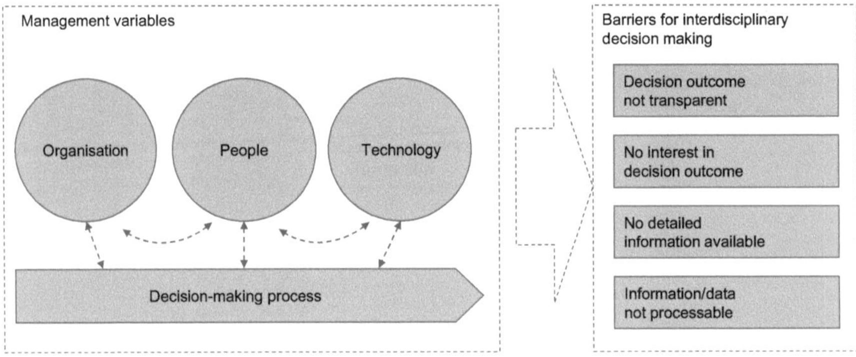

Fig. 3 Basic management variables that can influence interdisciplinary decision-making (based on Leavitt 1978)

Concerning the dimension *organization*, adequate structures and processes can facilitate interdisciplinary decision-making. Especially, interdisciplinary working groups, project teams, and the setup of an interdisciplinary project management have proven successful and lead to a cross-functional organization. Knowledge management has to be institutionalized to ensure that interdisciplinary knowledge is not the outcome of one-time projects and then easily lost in the process.

Regarding the *people* working in interdisciplinary contexts, almost no measures for interdisciplinary training and education do exist. Such measures, together with frequent workshops, are appropriate to facilitate an understanding of the interdisciplinary consequences of decisions. In addition, job rotation programs as well as the introduction of adequate incentive systems can further improve decision-making. Also, while technology today increasingly allows to exchange information across functional and organizational boundaries, people have to be educated and motivated to use such solutions.

Concerning *technology*, standards for integrated IT systems, systems' interfaces, and data can help to facilitate data interchange in interdisciplinary contexts. For example, currently the development of industry standards like VDA and ODETTE contributes to interdisciplinary integration. On the technological side, one of the biggest obstacles is the integration of decision variables from different methods and models from various disciplines. To overcome this obstacle, we developed a typology of decision variable interrelations which allows to identify adequate decision support solutions based on the interrelation type. A framework consisting of a typology of decision variable interrelations as well as a typology of decision support solutions will be presented in the following.

Development of a Framework for Characterizing Interrelations and Identifying Adequate Decision Support Solutions

Types of Decisions and Decision Interrelation Types in Production, Logistics, and Traffic

During the case study, we observed that decisions and therefore also the resulting interrelations are of different nature. While a decision which, e.g., concerns the production lot size can easily described by a numeric value, a decision upon the layout of a production facility or even concerning a strategy to enter a certain market is much more complex. In literature, there are various ways to structure or stereotype interrelations between decisions (Vidaillet 2012). Basically, decision variables can be distinguished into qualitative and quantitative decision variables. Quantitative variables can again be distinguished into atomic and composed (or complex) types (Chakhar and Mousseau 2004). Atomic decision variables represent the simplest form of a decision variable. They usually hold a single or multiple figures with their associated measurement units and may represent, e.g., coordinates, times, costs, or distances.

Composed decision variables are constructed through combining and interconnecting atomic decision variables. Composed decisions include the layout of a warehouse or production site as well as a transport route. These composed decisions can be formalized, e.g., by building a model, without the need of abstraction or simplification.

In contrast to composed decision variables, qualitative decision variables, such as the qualification or motivation of workers, cannot easily be composed from atomic variables. Modeling these variables is only possible by abstraction or simplification. This always leads to a loss of information that probably is required to make the *right* decision. Nine different types of interrelations by generating all permutations of the different decision variables are shown in Fig. 4.

Decision Support Solutions and Their Applicability in Interdisciplinary Contexts

Various disciplines are committed to research in the area of decision support solutions (Bohanec 2003), e.g., operations research, management science, decision science, or information management science. In those fields, different typologies or classifications do exist, which provide a better understanding of the roles and functions of different decision support solutions and help to align adequate decision support solutions to specific decision problems. Existing decision support solutions include decision support systems, business intelligence, optimization methods, and

independent variable \ dependent variable	qualitative	composed	atomic
qualitative	QQ	QC	QA
composed	CQ	CC	CA
atomic	AQ	AC	AA

interrelation types

Fig. 4 Different types of interrelations between decision variables

models as well as creativity methods. Due to the heterogeneity of the respective disciplines, there is no interdisciplinary approach that allows for classification or typecast of interdisciplinary decision support solutions. Hence, in order to derive requirements for interdisciplinary decision support, we derived a generally applicable decision support typology, which builds the fundament for further evaluation.

Based on the decision-making process, the decision makers' tasks can be broken down to the subtasks information management, solution finding, and communication. Information can be interpreted as a composed subset of purposive, context relevant data. This allows for interpretation and derivation of decisions in the given context (Sabherwal 2011). Capabilities of existing information management solutions are *collecting*, *filtering and transforming*, *data mining*, and *preparing and presenting* available information (Hansen 2008; Power 2002). Methods or models support the finding of an optimal solution for a given situation. Models encompass decision models, evaluation models, and description models (Bretzke 1980). Methods include creativity, structuration, and causality methods as well as heuristics and exact methods (Karnopp et al. 2012; Eiselt and Sandblom 2004; Bronner 1999; Bretzke 1980). Decision support solutions can also facilitate communication during the decision-making process (Breslin et al. 2010). While decision support solutions might only be capable of informing other decision makers of the decision outcome (unidirectional communication), they might also be implemented during the decision-making process (bidirectional communication). This should be done in order to gather feedback from other decision makers who are potentially affected by a decision. Possible capabilities of decision support solutions are shown in Fig. 5.

Type of decision support	*Capabilities*					
1. Information management	Collecting	Filtering & transforming	Data Mining/text mining	Preparing & presenting		*None*
2. Models	Decision model	Evaluation model	Description model			*None*
1. Methods	Creativity	Structuration	Causality	Heuristic	Exact	*None*
1. Communication	Unidirectional	Bidirectional				*None*

Fig. 5 Possible capabilities of decision support solutions

Derivation of Requirements for Interdisciplinary Decision Support

In the following chapter, requirements for the different types of decision support in interdisciplinary contexts are derived according to the different types of decision support and their various capabilities listed above.

Information Management

Concerning atomic and composed decision variables (relation types *AA, AC, CA, CC*), capabilities for transforming, filtering, and collecting data are sufficient. In the interdisciplinary context, adequate transformation rules for the measures in one system to measures in another system have to exist at an atomic level. In case of composed variables, the development of adequate transformation rules might hinder the implementation of interdisciplinary decision support. More complex tools have to be implemented, if qualitative decisions are concerned. If a qualitative decision triggers another decision (relation types *QA, QC, QQ*), adequate heuristics have to be implemented in order to transform qualitative into quantitative data. Data mining methods are an adequate solution in this case. If a qualitative decision is triggered by another decision (relation types *QQ, AQ, CQ*), methods for preparing and presenting existing information, which might be of atomic, composed, or qualitative nature, will build the basis for adequate decision support. Such interrelations, which include at least one qualitative decision variable, represent one of the biggest challenges in interdisciplinary decision-making due to the utilization of different terminology and the lack of knowledge concerning domain-specific adherences concerning decisions from foreign subsystems. This is emphasized by another result: most of the barriers observed in the case study included qualitative decision variables.

Models

On the level of atomic decision variables, interrelations can be easily modeled by implementing decision models (relation type *AA*). But in an interdisciplinary context, decision models can only be implemented, if existing information management solutions facilitate the transformation of atomic variables from one subsystem to another or if overall accepted performance indicators do exist. In case one of the decision variables is of composed nature (relation types *AC, CA, CC*), the implementation of a decision model might not be feasible due to the fact that the resulting (mathematical) problem might not be easily computed. In this case, the usage of evaluation models, which allow for manually or automatically evaluating various configurations, is more adequate. In the interdisciplinary context, decision support for interrelations between composed and atomic/composed decision variables is not only impeded due to the missing standards but also due to the fact that the available data from various domains like production and traffic cannot easily be linked, due to the micro–meso–macro gap (Sjoestedt 2005). In this case, existing models have to be enhanced in order to be able to translate data from one domain or level to another.

Evaluation models can also provide relevant information for qualitative decisions (relation types *AQ, CQ, QQ*), but in this case given information has to be interpreted by the decision maker himself. In case that the initial decision is of qualitative type (relation types *QA, QC, QQ*), description models can help to understand the nature of the interrelation and thus support the decision-making process.

Methods

In case of atomic variables (relation type *AA*), exact methods can easily compute and mathematically express the interrelation between two decision variables, as far as the information management solutions provide the methods to transform data between the various subsystems and domains. In case composed decision variables are also considered (relation types *AC, CA, CC*), heuristics might have to be implemented instead of exact methods due to the increasing complexity of the problem. Also, the same impediments account concerning interdisciplinary decision support as described above concerning decision models: adequate rules for translating data between the various domains have to be developed in order to provide meaningful heuristics and exact methods for interdisciplinary decision support. In cases where qualitative decision variables are concerned (relation types *QA, QC, QQ, AQ, CQ*), exact methods are not applicable. In such cases, heuristics, structuration methods, causality methods, or creativity methods can be implemented. In interdisciplinary contexts, those methods do not only support singular decisions but they help to enlarge the decision maker's overall knowledge of the interdisciplinary domains and hence also facilitate improved decision-making concerning future interdisciplinary decisions.

Communication

Our case study showed that if only atomic decision variables are concerned, the interrelations of decisions are better understood than in cases where also composed or qualitative variables are at stake. Hence in case of *AA*-type decision variables, the communication of the decision outcome to affected parties is sufficient. If also composed decision variables are part of an interrelation (relation types *AC, CA, CC*), the need for bidirectional communication should be further evaluated based on the degree of interdisciplinary knowledge of the decision makers involved. Especially if the interrelation between the two decision variables cannot be easily described, as it is often the case concerning interrelations between the micro- and the macro-level (Zuber et al. 2014), affected parties should be integrated early in the decision-making process. If qualitative decision variables are concerned (relation types *QA, QC, QQ, AQ, CQ*), potentially affected decision makers should always be included early in the decision-making process in order to control the decision outcome. In this case, decision support solutions facilitating bidirectional communication should be implemented.

In the following, we will derive the requirements for interdisciplinary decision support based on an exemplary interrelation comprehending the decision variables *production strategy* and *means of transport*.

Exemplary Discussion of the Decision Support Requirements

A significant development is the rise of flexibility in production paired with the reduction of lot sizes to react on current trends like increasing volatility. Those items are usually qualitatively defined by a company's *production strategy* which also interrelates with the composed decision variable *means of transport* from *outbound logistics*.

Since the decision variable production strategy is of qualitative type, in addition to the basic information management capabilities, advanced data mining or text mining technologies, which allow for interpreting and quantifying qualitative data, could be implemented in the area of information management. For example from a qualitative defined *production strategy,* lot sizes or required degree of flexibility could be extracted allowing to facilitate the understanding of the relationship with the decision variable *means of transport*. Concerning solution finding approaches, structuration methods in addition to descriptive models could be used in order to facilitate the understanding of possible effects of various stereotypes of production strategies. Broad interdisciplinary knowledge is not only required when designing adequate decision support solutions but should also be accessed during the decision-making process. This means, decision makers from *outbound logistics* should be consulted early in the decision-making process concerning the *production strategy* by using bidirectional communication tools.

Despite the remarks above, we observed in practice that interrelationships of the *QC* type are often not supported by decision support solutions. Due to a lack of adequate solutions, which would meet the described requirements, qualitative decisions are often implemented without caring about their impact on decisions on the composed or atomic level. Based on our observations, the same is applicable for all interrelations comprehending qualitative decision variables (relation types *AQ*, *CQ*, *QQ*, *QC*, *QA*).

Conclusion

The growing complexity of value networks will increase the need for integrated decision-making approaches in the future. Experiences from production, logistics, and traffic and transport show that interdependencies exist and need to be managed, that system-wide decision-making is not implemented in practice, and that there are four major barriers for interdisciplinary decision-making.

The derived barriers for interdisciplinary decision-making represent a first approach to identify and systemize the causes that detain companies from implementing interdisciplinary decision approaches that most likely would lead to more efficient and more effective value networks. Further, the provided systematization builds a solid foundation for developing adequate methods to overcome these barriers and to increase the quality of decision outcomes. Based on this foundation, measures could be identified that can support interdisciplinary decision-making processes.

In order to implement those measures, a deeper understanding of the decision variables is required. This is demonstrated in the area of technology by the development of a typology of decision variables and the identification of adequate decision support solutions. One of the biggest leanings of this approach is that due to ongoing standardisation efforts (e.g., industry standards) interrelations between decision variables on the atomic level can be quite well supported while decision support for interrelations including complex or qualitative variables is still not easy to implement. This is mainly the cause due to the micro–meso–macro gap.

Theoretical and Managerial Implications

As already outlined in this chapter, existing research in the area of interdisciplinary decision-making is either focused on general aspects of interdisciplinary decision-making processes or related to some niche application fields. Thus, this chapter represents a first approach to illuminate barriers of interdisciplinary decision-making and their reasons concerning decision-making in value networks. It also represents one of the few existing approaches to build up a general decision support typology allowing to characterize different decision support solutions.

The barriers identified in the chapter are usually not the subject of the discussions in decision science; even so through the listed barriers it became apparent that the human factor plays an important role in interdisciplinary decision-making. Evidently, the human factor has considerable potential to increase the efficiency of decision processes. The different requirements arising from the systems state spaces require different types of decision makers within the company. But these requirements are versatile and the dominance of qualitative and composed decision variables shows that purely technocratic and mechanistic decision-making aids are not easily applicable. Better solutions based on associated models could be developed, particularly for times of uncertainty or unstable environment. For academics it is necessary to understand boundaries, connection, and effects of specific decisions.

Due to ideas like Supply Chain Management or Supplier Relation Management, collaboration was no foreign term to the companies that were subject to our case study. Nonetheless, the applied case study proved that companies are often far away from a high degree of supply chain integration. Even simple decision processes, like for example the decision to use a certain type of container, are not coordinated even so this type of decision affects multiple parties among different companies, e.g., supplier, transport operator, and the OEM. Repacking the goods in this case does not only increase the process complexity thus causing a higher error rate and longer procession times but of course also increases the costs for at least one party due to additionally required resources. Also, having a specific decision problem at hand, practitioners often do not know what type of decision support solution to implement.

Having this in mind, the barriers for interdisciplinary decision-making identified in this chapter and the developed framework are of high value to practitioners. They profit by gaining knowledge about what might be the reasons for inefficiencies or complex processes in their companies or value networks and they can build on the derived typologies in order to implement the right decision support solutions.

Furthermore, the result implies that the human factor has considerable potential to increase the efficiency of processes. This happens when not only one's own process but also upstream and downstream steps are considered. It will be essential that employees develop an awareness of the implications of their actions for other parts of the process chain.

Research Limitations and Future Research Directions

This chapter presents a first idea on the barriers, which detain companies from making more integrated decisions, and presents measures, especially in the area of technology, which can support interdisciplinary decision-making. Concerning the case study's findings, numerous questions evolved during the interpretation of the case study's results:

- The case study showed that the lack of interdisciplinary decision-making has negative impacts, e.g., on supply chain performance. Performance impacts of interdisciplinary decision-making should be identified in order to be able to focus integration efforts to the most rewarding areas.
- A more differentiated set of integration measures has to be developed that allows to effectively address the different barriers.
- In order to further derive requirements for decision support, the interrelations on the decision level in the interdisciplinary system of production, logistics, and traffic have to be further investigated. For example, methods from (social) network analyses can be implemented in order to further describe the role of the various subsystems and the different types of decision variables in order to derive decision support requirements. A first approach is presented in Zuber et al. (2014).
- Interrelations showing barriers are dominated by qualitative and composed decision variables. Overcoming the so-called micro–meso–macro gap in order to understand the relations between those levels, e.g., the impact of decisions concerning the transport infrastructure on delivery schedules in logistics, is essential to overcome the identified barriers.
- Our literature review revealed that interdisciplinary research concerning the derivation of problem-specific decision support is sparse and often outdated (e.g., Gorry and Morton 1971). Our approach can build a fundament for future frameworks to enhance the development of decision support solutions.

The identified research directions do reveal that with our research we could only scratch the tip of the iceberg. With the identified barriers and the derived typologies for interrelations and decision support solutions, we provide a solid foundation for further research on integrated decision-making.

References

Bell, D. E., Raiffa, H., & Tversky, A. (1988). Descriptive, normative, and prescriptive interactions in decision making. In D. E. Bell, H. Raiffa, & A. Tversky (Eds.), *Decision making: Descriptive, normative, and prescriptive interactions* (Vol. 1, pp. 9–32). Cambridge: Cambridge University press.

Blackhurst, J., Wu, T., & O'Grady, P. (2005). PCDM: A decision support modeling methodology for supply chain, product and process design decisions. *Journal of Operations Management, 23*(3), 325–343.

Blecker, T., & Kersten, W. (2006). *Managing risks in supply chains: How to build reliable collaboration in logistics*. Berlin: Erich Schmidt Verlag GmbH.

Bohanec, M. (2003). Decision support. In D. Mladenić, N. Lavrač, M. Bohanec, & S. Moyle (Eds.), *Data mining and decision support* (pp. 23–35). Boston, MA: Springer.

Boltze, M., Rühl, F., Berbner, U., Friedrich, H. (2015). *The interdisciplinary decision map – A reference model for production, logistics and traffic*. In TRB 94th Annual Meeting, 11-15 Jan 2015, Washington DC, USA.

Breslin, J. G., Burg, T. N., Kim, H.-G., Raftery, T., & Schmidt, J.-H. (2010). Recent trends and developments. In *Social software: International conferences on social software*, BlogTalk 2008, Cork, Ireland, March 3–4, 2008, and BlogTalk 2009, Jeju Island, South Korea, September 15–16, 2009. Revised Selected Papers. Springer.

Bretzke, W.-R. (1980). *Der Problembezug von Entscheidungsmodellen*. Tuebingen: Mohr Siebeck.

Bronner, R. (1999). *Planung und Entscheidung: [Grundlagen – Methoden – Fallstudien]*. Wien: Oldenbourg.

Chakhar, S., & Mousseau, V. (2004). Towards a typology of spatial decision problems. *Annales Du Lamsade N 2 Juin 2004*, 125.

Craighead, C., Blackhurst, J., Rungtusanatham, M., & Handfield, R. (2007). The severity of supply chain disruptions. Design characteristics and mitigation capabilities. *Decision Science, 38*(1), 131–156.

Dillon, S. M. (1998). *Descriptive decision making: Comparing theory with practice*. Unpublished manuscript, Department of Management Systems, University of Waikato.

Eiselt, H. A., & Sandblom, C.-L. (2004). *Decision analysis, location models, and scheduling problems*. Berlin: Springer.

Eisenhardt, K. (1989). Building theories from case study research. *The Academy of Management Review, 14*, 532–550.

Elwyn, G., & Edwards, A. (2009). *Shared decision-making in health care* (2nd ed.). Oxford: Oxford University Press.

Frankel, R., Bolumole, Y. A., Eltantawy, R. A., Paulraj, A., & Gundlach, G. T. (2008). The domain and scope of SCM's foundational disciplines—Insights and issues to advance research. *Journal of Business Logistics, 29*(1), 1–30.

Gligor, D. M., Holcomb, M. C., & Stank, T. P. (2013). A multidisciplinary approach to supply chain agility: Conceptualization and scale development. *Journal of Business Logistics, 34*(2), 94–108.

Gorry, G. A., & Morton, M. S. S. (1971). A framework for management. Information systems. *Sloan Management Review, 1971*, 21–36.

Hansen, T. (2008). *RFID-gestützte Produktempfehlung im stationären Einzelhandel*. Berlin: Logos Berlin.

Harrison, E. F. (1993). Interdisciplinary models of decision-making. *Management Decision, 31*(8), 27–33.

Huber, G. P. (1980). *Managerial decision making*. Glenview, IL: Scott, Foresman.

Karnopp, D. C., Margolis, D. L., & Rosenberg, R. C. (2012). *System dynamics: Modeling, simulation, and control of mechatronic systems*. Hoboken: Wiley.

Koutsoukis, N.-S., & Mitra, G. (2003). *Decision modelling and information systems: The information value chain*. Norwell, MA: Kluwer Academic Publishers.

Lambert, D. M., García-Dastugue, S. J., & Croxton, K. L. (2005). An evaluation of process-oriented supply chain management frameworks. *Journal of business Logistics, 26*(1), 25–51.

Laux, H. (2012). *Entscheidungstheorie* (8th ed.). Berlin: Springer.

Leavitt, H. J. (1965). Applied organizational change in industry: Structural, technological, and humanistic approaches. In J. G. March (Ed.), *Handbook of organizations* (pp. 1144–1170). Chicago, IL: Rand McNally.

Leavitt, H. J. (1978). *Managerial psychology. An introduction to individuals, Pairs, and groups in organizations* (4th ed.). Chicago, IL: The University of Chicago Press.

Mezias, J. M., & Starbuck, W. H. (2008). Decision-making within accurate, unreliable data. In G. P. Hodgkinson & W. H. Starbuck (Eds.), *The Oxford handbook of organizational decision-making* (pp. 76–96). New York: Oxford University Press.

Pfohl, H.-Chr. (1977). *Problemorientierte Entscheidungsfindung in Organisationen*. Berlin.

Pfohl, H.-Chr., Berbner, U., & Ehrenhoefer, M. (2012a). Fundamental problematic issues in interdisciplinary decision-making: An integrated view on production, logistics, and traffic and transport. In RIRL (Ed.), *Actes des rencontres RIRL 2012 Proceedings Gestion de la chaine logistique Supply Chain Management.*

Pfohl, H.-Chr., Ehrenhöfer, M., & Zuber, C. (2012b). Dynamic and seamless integration of production, logistics and traffic/transport (Dynamo PLV)—Challenges of an interdisciplinary research project. *Journal of Management, 11*, 81–95.

Power, D. J. (2002). *Decision support systems: Concepts and resources for managers.* Westport, CT: Greenwood Publishing Group.

Rahman, N., & De Feis, G. L. (2009). Strategic decision-making: Models and methods in the face of complexity and time pressure. *Journal of General Management, 35*(2), 43–59.

Sabherwal, R. (2011). *Business intelligence: Practices, technologies, and management.* Hoboken, NJ: Wiley.

Sanders, N. R., Zacharia, Z. G., & Fugate, B. S. (2013). The interdisciplinary future of supply chain management research: Interdisciplinary future of SCM research. *Decision Sciences, 44*(3), 413–429.

Schramm, W. (1971). *Notes on case studies of instructional media projects.* Working Paper for the Academy of Educational Development. Washington, Educational Resources Information Center.

Simon, H. A. (1997). *Administrative behavior. A study of decision-making processes in administrative organizations.* New York: Free Press.

Sjoestedt, L. (2005). A conceptual framework for analysing policy maker's and industry roles and perspectives in the context of sustainable goods transportation. In P. Rietveld & R. R. Stoud (Eds.), *Barriers to sustainable transport: Institutions, regulation and sustainability* (pp. 198–222). London: Routledge.

Straube, F., & Pfohl, H.-Chr. (2008). Trends and strategies in logistics. Global networks in an era of change. Environment, security, internationalisation, people. Bremen.

Strausa, S. E., Tetroeb, J. M., & Grahamb, I. D. (2011). Knowledge translation is the use of knowledge in health care decision-making. *Journal of Clinical Epidemiology, 64*(1), 6–10.

Vidaillet, B. (2012). When 'decision outcomes' are not the outcomes of decision. In G. P. Hodgkinson & W. H. Starbuck (Eds.), *The Oxford handbook of organizational decision making.* Oxford: Oxford University Press.

Yin, R. (2009). *Case study research: Design and methods* (4th ed.). Thousand Oaks, CA: Sage.

Zuber, C., Pfohl, H.-C., & Berbner, U. (2014). Integrating domains in supply chains—Development of requirements for interdisciplinary decision support based on an integrative framework of production, logistics, and traffic. In *5th Annual Conference of the European Decision Sciences Institute*, Denmark, Kolding.

The Interdisciplinary Decision Map: A Reference Model for Production, Logistics and Traffic

Manfred Boltze, Frederik Rühl, Ulrich Berbner, and Hanno Friedrich

Introduction

There has always been a strong interference between economic activities and the transport system (Manheim 1979) and especially decisions in supply chains do not usually stand for their own but are highly interrelated (Vidaillet 2008). This holds especially true due to newer supply chain concepts such as just in sequence and developments towards lean supply chains, which have increased supply chain complexity over the past years (Blecker and Kersten 2006; Craighead et al. 2007). Accordingly, decisions by one decision-maker impact other decision-maker's actions. For example, decisions on the production programme indirectly determine the need for transport. The transport service provider fulfils this demand and takes decisions such as route choice, which have an impact on the traffic system. Vice versa, traffic measures such as truck bans or HGV tolls can influence

M. Boltze (✉) • F. Rühl
Chair of Transport Planning and Traffic Engineering, Technische Universität Darmstadt, Otto-Berndt-Straße 2, 64287 Darmstadt, Germany
e-mail: boltze@verkehr.tu-darmstadt.de; ruehl@verkehr.tu-darmstadt.de

U. Berbner
Supply Chain and Networkmanagement, Technische Universität Darmstadt, Hochschulstraße 1, 64289 Darmstadt, Germany
e-mail: berbner@bwl.tu-darmstadt.de

H. Friedrich
Kühne Logistics University – KLU, Wissenschaftliche Hochschule für Logistik und Unternehmensführung, Grosser Grasbrook 17, 20457 Hamburg, Germany
e-mail: hanno.friedrich@the-klu.org

E. Abele et al. (eds.), *Dynamic and Seamless Integration of Production, Logistics and Traffic*, DOI 10.1007/978-3-319-41097-5_3

production processes and related transports in multiple ways. In case of implementation, companies will have to adapt their processes to new conditions to avoid problems or raising costs. Notwithstanding the above, both companies and transport-related public authorities rarely take other participant's needs into account when making decisions (Anderson et al. 2005; Stathopoulos et al. 2012). However, it is desirable to optimise decision-making processes not only within a single discipline (production, logistics, traffic) but to realise a coordinated decision that takes the objectives of various probably affected decision-makers into account. Derived from decision theory, decision processes including decision-makers from different disciplines will be further addressed as interdisciplinary decision-making (Pfohl et al. 2012).

The reasons for the lack of interdisciplinary decision-making are numerous and can be attributed to the fields human, structure and technology (Pfohl et al. 2012). All those factors can either facilitate or impede interdisciplinary decision-making. For example, decision-makers who have experienced an interdisciplinary education are more likely to consider interdisciplinary effects of their decisions than others. Also, interdisciplinary structures such as comprehensive working groups can facilitate interdisciplinary decision-making. Nonetheless, in order to increase effectiveness and efficiency of decisions, well-educated employees and adequate structures are not enough: methods and instruments (technologies) are needed in order to support decision-makers especially in case of rather complex decision problems.

Suitable approach in this context is the use of reference models. Reference models structure specific vocabulary, decision spaces and processes and, thus, shape a common and consistent understanding among decision-makers within one discipline (Zuber et al. 2015). In the interdisciplinary context, only few such models do exist. One example is the Supply Chain Operations Reference Model (SCOR), which includes decision variables and processes from the logistics and production domain (Supply Chain Council 2010). However, there is no model describing interrelations between production, logistics and traffic.

Hence, this chapter aims to propose a reference model which, first, allows for a common understanding among decision-makers for the interdisciplinary domain of production, logistics and traffic. Second, the model shall allow decision-makers from that interdisciplinary domain to describe and understand impacts of own decisions across the disciplines. Thus, for instance, potential users of the presented model may be decision-makers in planning agencies who assess the effectiveness of city logistics measures as well as companies who need to estimate impacts of decisions made by their supply chain partners on their own business processes.

The development of such an integrated model is complex due to different and sometimes contrary requirements: On the one hand, claiming completeness requires the smallest details. The model must support aggregated as well as fine-grained impact analyses, since decisions are made on different levels of granularity. For example, the impact analysis for adaptive traffic signal control as short-term decision has other requirements than the one for analysing impacts of clustering firms within land use management. On the other hand, subsequent users require usability and low complexity. A transport planner pondering over a decision for a truck ban, for example, is only interested in significant impacts, not in eventualities.

Consequently, since comprehensibility and usability suffer from a too excessive degree of complexity, the model must be limited to the information necessary for the user. Since the model shall be used in different disciplines, flexibility to determine the level of detail in different parts of the framework is required.

As extension of Rühl et al. (2013), the reference model presented in this chapter provides a structural framework for mapping decisions from the disciplines production, logistics and traffic as well as a modelling language for describing interrelations between those decision variables. Since the developed reference model allows for mapping individual decisions and for locating interrelations of those decisions, it will be further referred to it as the Interdisciplinary Decision Map (IDM).

In the following, an overview is given concerning interdisciplinary decision-making. Based on underlying theories, requirements for the IDM as an adequate reference model are derived. Subsequently, it is shown how the IDM was developed based on theory and a case study approach and how it was validated using interdisciplinary scenarios. Furthermore, the IDM's application is illustrated by the example of the introduction of HGV tolls and its impacts on business processes. The chapter closes with findings and an outlook on future extendibility and utilisation of the developed reference model.

Understanding Interdisciplinary Decisions and Their Interrelations

Existing Reference Models in the Context of Production, Logistics and Traffic

Challenges for interdisciplinary reference models are the differing research goals and the diverse cultural imprints of the considered fields of research, but also a lack of knowledge concerning the opposite research fields and the existing terminology (Sanders et al. 2013). This holds true for all practical circumstances where decisions have to be made in an interdisciplinary context (Pfohl et al. 2012, 2013).

Various research fields and research disciplines utilise so-called reference models or ontologies to facilitate a common understanding of their domain. While reference models usually describe relevant elements with their characteristics and provide a common language for describing interrelations, ontologies also define hierarchical or constitutional relations between their elements (Bera et al. 2011). In literature, reference models are described as "abstract frameworks for identifying and defining concepts, as well as significant relationships among the entities of a domain" (Bahm et al. 2011). Examples are the above-mentioned SCOR model (Supply Chain Council 2010) or GenCLOn, an ontology that has been developed to model stakeholders and their objectives in city logistics (Anand et al. 2012).

Reference models are always designed to fulfil a certain purpose; e.g. the SCOR model's purpose is to support supply chain experts with a standardised terminology that helps them to achieve a holistic supply chain understanding and to develop integrated supply chain models (Poluha 2007). While there are already recent modelling approaches that integrate limited aspects from production, logistics and traffic (e.g. Clausen et al 2008; Roorda et al. 2010), most approaches do not represent general reference models. Instead, they are usually highly abstracted and build by researchers from a certain discipline to fulfil a very narrow task (e.g. optimised tour planning). Describing such a reference model that enables to analyse the various interdependencies from an interdisciplinary perspective, i.e. including researchers and practitioners from production, logistics and traffic, is goal of this chapter.

Systems Theory and Interdisciplinary Decisions

Systems theory and decision theory can be utilised in order to build the theoretical foundation for such a general reference model, while Porter's value chain model (Porter 1985) can serve as a structural framework.

Systems theory is a universal approach (Luhmann 2011), where a system can be seen as interlinked items or subsystems, which are separated from their environment (Bertalanffy 1976; Luhmann 2011). In case of open systems, the system's border can be seen as a permeable membrane, which allows the system to interfere with its environment (Thom 2008). Those interferences can be described by input and output relations, which e.g. might represent flows of information or materials (Roggisch and Wyssuek 2002). While those interferences, which are also referred to as structural linkages (Krause 2005), do not allow a system to directly influence items in another system, external influences can enforce internal changes (Luhmann 2011). Structural linkages can be differentiated into loose couplings and strict couplings. In case of the latter, the information received by one system (input) equals the information sent by another system (output) (Krause 2005). In practice, this only holds true for limited cases. Usually, information has to be translated on its way from one system to another.

Decisions or decision outcomes can provide a structural linkage between various subsystems, since decisions are highly interrelated in the interdisciplinary context (Vidaillet 2008). Decisions in one system are usually triggered by a change, which is either caused by the environment or by another decision with relevance to the decision-maker's domain.

Speaking in terms of the decision-making process, the last step of the inducing process (making decision/implementation) is directly linked with the first step of the triggered decision-making process (need for action/situation analysis/intelligence) (see Fig. 1). A decision's impacts are usually monitored by various performance indicators. General indicators, which are used broadly over various disciplines, are quality, flexibility, cost and time (Wang and Masini 2009). These indicators have always to be seen in their relevant context; e.g. costs in logistics might have a different meaning from costs in production. For examples of

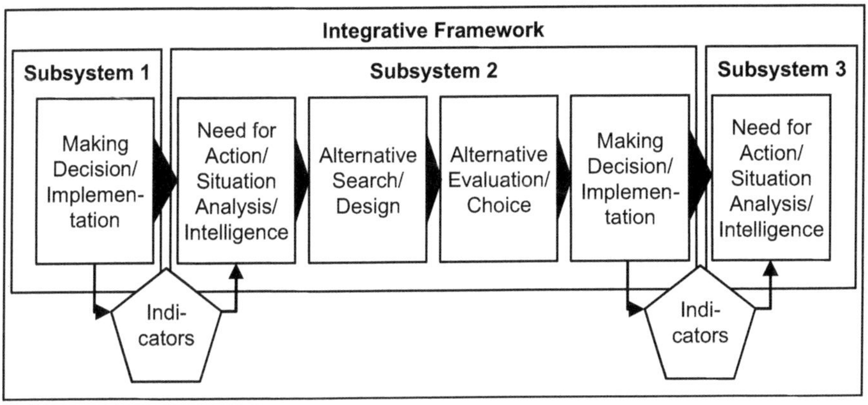

Fig. 1 Interdisciplinary decisions in the integrative framework of production, logistics and traffic (based on Zuber et al. 2015)

performance indicators from the different domains, see (Martínez and Pérez 2001; Miranda and Rodrigues da Silva 2012). Hence, as stated above, to describe the structural coupling between two systems, multiple translation processes are required: A decision will be measured by indicators (translation 1), the indicator in one system will be translated to another system (translation 2) and then this indicator will be interpreted (translation 3) and can trigger other decisions.

The role of decisions concerning the structural linkage between various subsystems has been studied in the research project Dynamo PLV (Pfohl et al. 2013). Actors studied in Dynamo PLV comprise decision-makers from public authorities (relevant for the traffic domain) and companies (relevant for the production and logistics domain). Companies are not merely split up into production and logistics departments, but rather build on a more complex and heterogeneous structure, which Porter (1985) tried to reflect by his interpretation of the value chain. The research in Dynamo PLV has revealed that e.g. departments like production, sourcing or inbound logistics do not share the same decision space nor objectives even so decisions made in one department often affect decisions in another (Pfohl et al. 2012). Zuber et al. (2015) provide further evidence that the model of Porter adequately reflects the domains and decision spaces of decision-makers in production, logistics and traffic by applying methods of network analysis to an interdisciplinary decision network. Also, Porter's model is well established and well known in all the disciplines addressed by the framework.

A decision-maker's decision space is tightly bound to his decision domain and thus defined by the function he is working in (Pfohl et al. 2012). There are different approaches to structure decisions and decision spaces; one of the most common is the so-called Leavitt Diamond (Maier et al. 2005). Leavitt categorises decisions into human, structure and technology (March 1965). Decisions in the area human can e.g. address staff qualification measures, decisions in the area structure might affect processes or the organisational structure, and decisions in the area technology include the introduction of new technological equipment, e.g. new IT systems.

Requirements for a Reference Model in the Context of Production, Logistics and Traffic

Based on the findings derived from theory, the following requirements for a comprehensive reference model for production, logistics and traffic can be summarised:

- Decision-makers often lack the understanding of the decision variables from other disciplines. Hence, the reference model should include a structural framework or decision map, which gives an overview of the existing subsystems and the decision-makers' decision spaces.
- In order to facilitate a common and easy understanding of the framework, it should be aligned to an existing model, which is already used in research and practice and well known in the disciplines production, logistics, and traffic. Hence, the reference model should adapt decision domains as functional units of the company as described by Porter (1985). Since traffic is not in the scope of Porter's model, it needs to be implemented as a separate decision domain/subsystem.
- In order to enhance the understanding of the substantial number of possible decision variables, they should be grouped by using a common structure. Hence, the reference model should adapt Leavitt's Diamond by grouping decision variables into the fields human, structure and technology.
- The interdisciplinary impacts of decisions are not well understood. Hence, the reference model should provide a method to describe structural linkages between two or more systems.

Methodology

In order to develop the IDM, a stepwise research approach was implemented.

In the first step, based on the theoretical findings presented above, a structure was developed to map the decision spaces of the different systems. The structure heavily leans on Porter's value chain while it also allows to distinguish between decision variables and performance indicators. Furthermore, the structure enables to group decision variables by the dimensions human, structure and technology. Also, a modelling language was developed that allows a systematical description of interrelations between decision variables. By using this modelling language, the impacts of decisions can be visualised as impact chains.

In the second step, the reference model was filled with actual decision variables from production, logistics and traffic based on literature review and case studies with industrial companies from the mechanical engineering and automotive industry, several transport service providers as well as transport-related public authorities.

Finally, the developed framework was validated by defining 20 scenarios which deal with various disciplinary decisions and their interdisciplinary dependencies, i.e. the modification of the means of transport in a certain production process, the adjustment of a distribution network or the introduction of HGV tolls. In doing so, the collection of decision variables and indicators was consolidated, a variety of interdisciplinary dependencies were added and significant interrelations between the given decision variables could be identified.

IDM for Production, Logistics and Traffic

Elements of the IDM

The IDM consists of three elements: The first element is the structural framework, which facilitates the collection and location of decision variables in the interdisciplinary decision space. The second element is the modelling language, which can be used to describe interrelations. The third element is the empirical data comprising decision variables, performance indicators and actual interrelations. The IDM and its constituting elements are shown in Fig. 2.

Structural Framework

Figure 3 shows the structural framework with its two-layered design, including decision variables on the first layer and indicators on the second.

The separation into decision variables and indicators follows the idea stated above that decisions themselves are often invisible, especially to decision-makers from other subsystems. Hence, decision outcomes are usually controlled or communicated by (measurable) performance indicators. While decision variables can be directly influenced by decisions (e.g. production strategy), indicators reflect measurable observations (e.g. produced pieces/minute). Often, a decision made does not only affect a single but multiple performance indicators.

The IDM's structure is aligned towards Porter's value chain (Porter 1985). Based on that, the considered corporate subsystems—purchasing, inbound logistics, production, intra-logistics, outbound logistics and sales—are embedded into the freight transport system (see upper part of Fig. 3).

To ensure comparability and, hence, to enable analyses of interrelations of decisions, Leavitt's diamond (Maier et al. 2005) is used to structure decisions within the subsystems. Consequently, the IDM allows the analysis of impacts of decisions related to people, structure or technology assuming a fixed task of a system given by its definition.

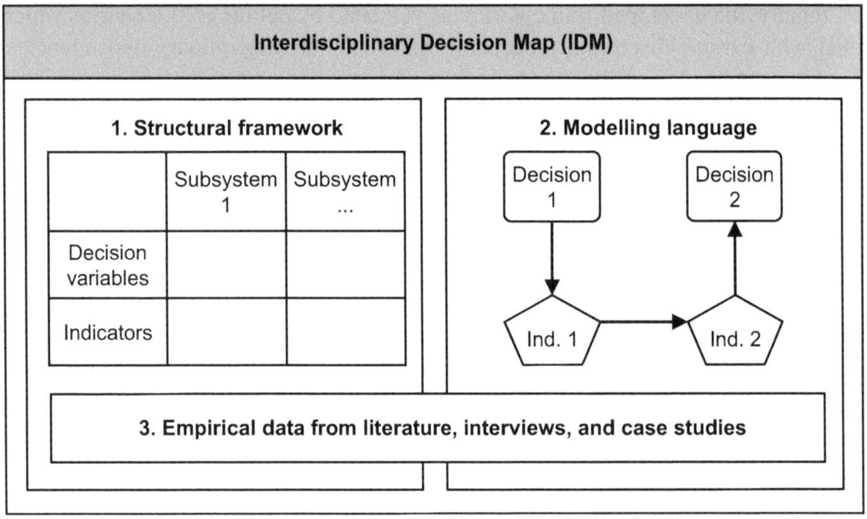

Fig. 2 Elements of the interdisciplinary decision map

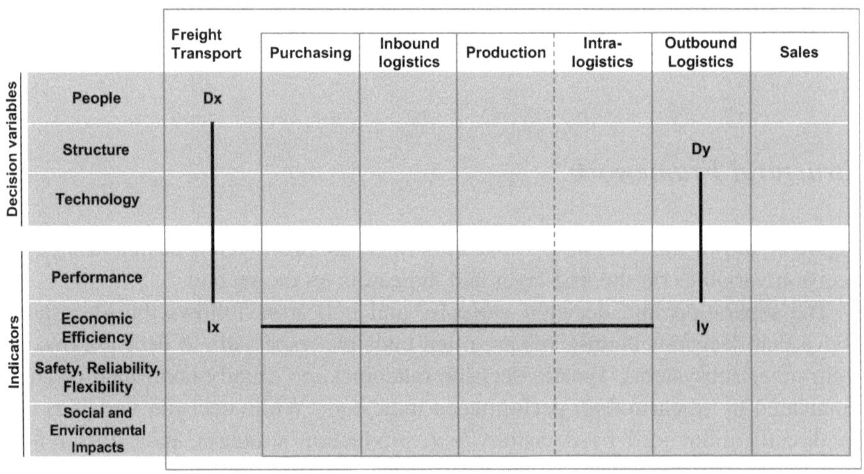

Fig. 3 Structural framework of the interdisciplinary decision

Furthermore, performance indicators are used to be able to evaluate decision's impacts. For this purpose, an indicator layer with quantifiable indicators for each system is integrated (lower part of Fig. 3). Similar to the layer of decision variables, also for the second layer a mutual structuring is needed. For this, subcategories 'performance' (e.g. capacities for production resources), 'economic efficiency' (e.g. cost, revenues), 'safety, reliability, flexibility' and 'social and environmental impacts' are suggested, following the four major goals of transport planning (Boltze et al. 1994).

Modelling Language

A modelling language allows for describing the impacts and interrelationship between two decisions. As previously mentioned, the modification of a decision variable will always have an impact on at least one performance indicator. If not, the decision would either be completely irrelevant and could be excluded from the IDM, or the IDM would be missing a relevant performance indicator to allow for measuring the decision's results. The subsequent change of a performance indicator might be observed either by decision-makers in the same system or by decision-makers in other systems triggering follow-up decisions. In case of the latter, performance variables of one system are often not transparent to decision-makers from another system. In those cases, often a structural linkage between multiple performance indicators can be observed. For example in practice, the indicator production lot size is usually directly linked to the indicator production cycle time. Taking this fact into account, a grammar to describe interrelations ($\rightarrow$) between decision variables D between two systems S including various performance indicators P looks like the following:

$$G: \ DS1 \rightarrow PS1 \rightarrow PS2 \rightarrow DS2$$

This grammar implies that decision $DS1$ influences performance indicator $PS1$ (called decision effect). The performance indicator $PS1$ might not be directly observable by a decision-maker from another system, but a structural link between $PS1$ and the observable indicator $PS2$ (called cause effect) might exist. When a change of $PS2$ is observed, a decision-maker might react by making another decision $DS2$ (called feedback effect).

Of course, there are some cases, where an adjacent decision-maker can directly observe a performance indicator $PS1$. In order to keep up with the developed structure of the modelling language, we suggest to use Grammar G^* to describe these types of interrelations:

$$G^*: \ DS1 \rightarrow PS1 \rightarrow PS1 \rightarrow DS2$$

Following this grammatical rule or modelling language, interrelations can be easily visualised inside the IDM, as illustrated in Fig. 3. Combining multiple interrelations allows for modelling interdisciplinary impact chains.

As shown in Fig. 3, displaying interrelations inside the IDM is quite intuitive. However, this type of visualisation is only feasible if the number of interrelations is limited. To allow modelling of extensive impact chains, graphical modelling elements and a tree-like modelling structure were developed to extend the modelling language (see Fig. 4).

The modelling language together with the developed structural framework provides a solid foundation for thoroughly analysing interrelations: First, it allows a clear illustration of the very complex topic. Second, due to its tree structure,

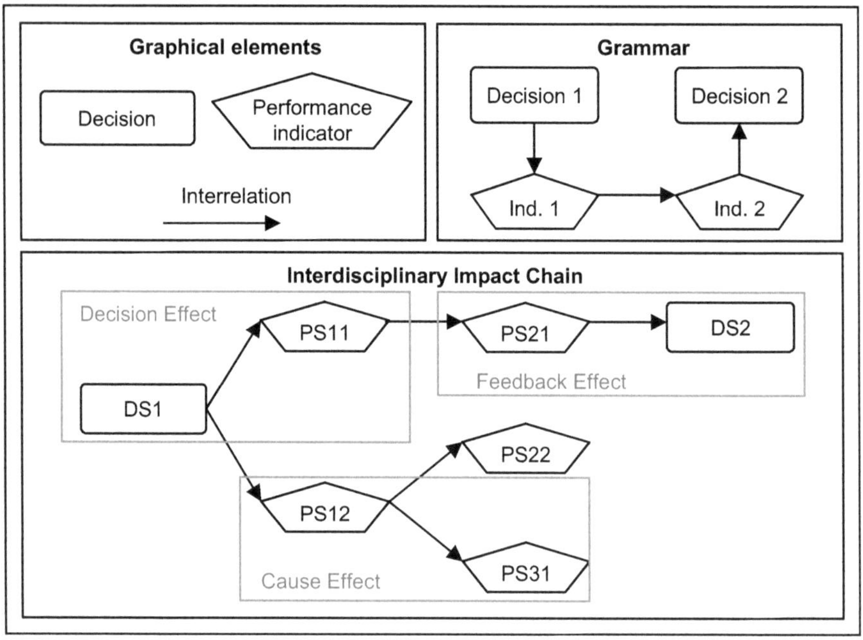

Fig. 4 Modelling language for modelling interdisciplinary impact chains

changes within the overall system caused by a decision can be analysed step by step. Third, by extending the graph with additional paths, the level of detail can be increased any time.

Within one subsystem, there may be decision effects, cause effects or feedback effects. Cross-system impacts are solely cause effects. Figure 4 illustrates this: Decision *DS*1 influences the indicators *PS*11 and *PS*12 (decision effect). As a consequence, indicators from other systems (*PS*21, *PS*22 and *PS*31) change as well (cause effect). Due to this, new decisions are made (*DS*2) (feedback effect).

Empirical Data: Decision Variables, Performance Indicators and Their Interrelations

As mentioned above, the IDM was initially filled with decision variables, indicators and the interrelationships by means of literature review and case studies. Subsequently, the impacts of disciplinary decisions from different domains were analysed in 20 defined scenarios (freight transport: 4; Purchasing: 2; Production/Intralogistics: 8; Outbound logistics: 6). An exemplary impact analysis for the introduction of HGV tolls is presented below.

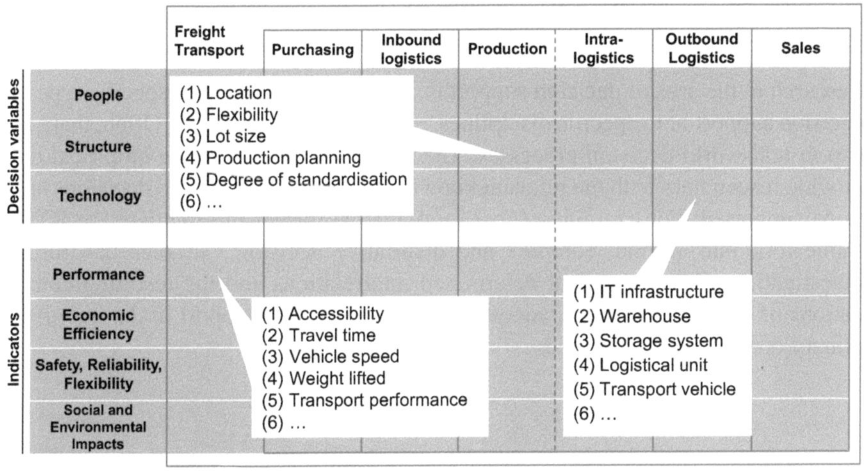

Fig. 5 Examples for decisions and indicators of the interdisciplinary decision map

In Fig. 5, the IDM is filled with examples of decision variables and indicators to illustrate its broad scope. A comprehensive list of all decision variables and indicators of the IDM is available from the authors.

Application Areas of the IDM

After its development, the IDM was already employed in various contexts during the research project Dynamo PLV in order to facilitate the follow-up on various research questions. A selection from these applications will be outlined in the following.

Impact Analysis and Checklists

Being able to change freight actors' habits in the context of freight transport demand management requires a profound knowledge about traffic measures' impacts on them (see Rühl and Boltze 2016). As mentioned above, impact analyses for traffic measures often neglect consequences for production and logistics, inter alia just due to the fact that there is a lack of certain knowledge in transport authorities. Accordingly, the IDM has been used for impact analyses of traffic measures to be able to highlight the possible range of impacts. Based on that, decision-making on sustainable traffic measures can be improved by developing checklists including critical interrelations to support decision-makers.

Decision Support Development

Research in the area of decision support is usually tightly bound to specific types of decision support or to specific disciplines, e.g. information systems. Often, there is a gap in real-world decision processes. The IDM and the underlying empirical data provide researchers with the relevant knowledge on how interrelated decisions have to be supported. For example, Zuber et al. (2015) further differentiate the IDM's framework into atomic, complex and qualitative decision variables. By further investigation of the empirical determined interrelations and the resulting combinations of decision variables, adequate tools for decision support in the interdisciplinary context can be derived.

Network Analysis

The developed modelling language allows for building extensive interdisciplinary impact chains which can easily be transformed into graphs where nodes represent decision variables or performance indicators and edges represent various types of interrelations, namely decision effects, cause effects and feedback effects. The structure of those graphs can then be further analysed by applying methods from graph theory/network analysis. For example, Zuber et al. (2015) apply various methods from network analysis (e.g. metrics like degree centrality) to the IDM in order to derive interrelation types and to allocate adequate types of decision support solutions.

Using the IDM to Analyse Impacts of the Introduction of HGV Tolls

Status Quo of HGV Toll Implementation

Heavy goods vehicle (HGV) tolls are strongly debated throughout politics, society and business. More and more countries, especially in the EU, are implementing HGV tolls. From the consideration to let users pay for infrastructure and ecological costs to the belief that the national transport business has to be protected against foreign competitors by road charging, there are a lot of different reasons for implementing HGV tolls. Manifold different HGV toll systems are in use. They differ regarding pricing parameters (time-based vs. distance-based) or system architecture (open vs. closed system). Descriptions of existing toll systems can be found in (McKinnon 2006; Broaddus and Gertz 2008; European Parliament 2008; Conway and Walton 2009).

Economic Impacts

Various macroeconomic impacts of the introduction of a toll system have been reported: slight consumer price increases due to higher transports costs (Doll and Schaffer 2007; Vasallo and López 2010), marginal employment effects (Doll and Schaffer 2007), a negative impact on accessibility (Condeço-Melhorado et al. 2011) and positive impacts on pollution emissions due to induced fleet modernisation (Musso and Rothengatter 2013).

Individual actors along the supply chain, i.e. carriers, logistics service providers as well as manufacturers, have to deal with toll-caused price increases to different degrees. The transport logistics sector has to deal with significant price increases up to 15 % while there are only marginal price effects on other industry sectors (Doll and Schaffer 2007; Vasallo and López 2010). Logistics costs in other industry sectors are usually less than 5 % of complete turnover. In the case of logistics service providers, they stand for roughly three-quarters (Einbock 2006). Hence, being worst affected, especially the transport sector seeks for alternatives to avoid or at least minimise cost increases.

Possible Reactions of Supply Chain Actors

According to Friedrich (2010), the business decisions leading to freight transport demand are distinguishable into rather long-term strategic decisions (e.g. business location choice) and short-term decisions regarding the operations (e.g. dispatching choices). It is assumed that on each of these hierarchical levels, actors may take action to react on HGV tolls. Thus, such decisions can be made either by the manufacturer, the logistics service provider and/or the carrier. While the manufacturer can react with a change of business location choice or sourcing choice over the long term, the logistics service provider has the competence for mid-term decisions as altering the transport chains and warehousing locations. Since the toll costs firstly occur for the carrier, he will take short-term actions concerning its operations (e.g. dispatching choices).

Academia has already paid attention to toll-caused decisions of enterprises and observable impacts within the traffic system. While research in logistics has addressed the problem from the user's perspective, in traffic engineering sciences impacts are examined on a more aggregated level. However, only Einbock (2006) addresses decisions from all different supply chain actors. European Parliament (2008) dwells on impacts on complete logistics systems, including the LSPs' and carriers' decisions. The focus in other references is restricted to the carrier's decisions (Gustafsson et al. 2006; European Parliament 2008; Hensher and Puckett 2008; Link 2008). Usually, manufacturer's decisions are not considered.

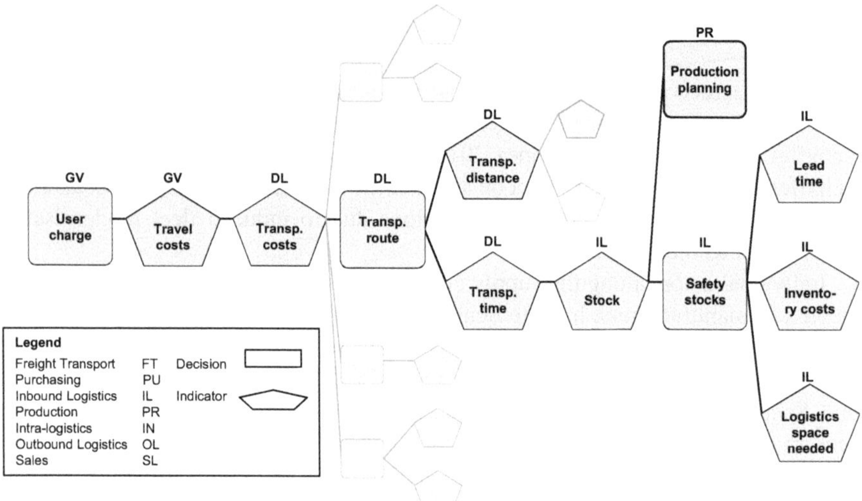

Fig. 6 Decision tree for the introduction of HGV tolls

Impact Analysis with the IDM

As previously mentioned, there are various possible reactions on an HGV toll introduction. Due to the complexity of the tree, in the following it is exemplarily dwelled on the carrier's route choice by stepwise discussing decisions and their impacts on production, logistics and traffic (see Fig. 6).

- The decision in the freight transport system to introduce HGV tolls as a measure for user charging leads to an increase of travel costs and, consequently, to increasing transport costs (indicator of logistics). To ensure its already tight margin, the logistics company has several opportunities to react such as to adapt the routing or to consider the modal choice. Assuming that the carrier adapts the routing, the indicators for transport distance and transport time may change.
- A change in transport time may result in a declining intra-logistics indicator measuring safety stocks, which in turn necessitates a decision on stocks, e.g. building up or increasing safety stocks.
- In the short run, changes in production planning and scheduling can cushion the effects of declining stocks on capacity utilisation. In the long run, a decision on increasing safety stocks influences indicators for lead time, inventory costs and logistics space needed, whose changes again trigger new decisions with implications for production, logistics and traffic.

Conclusions and Outlook

Due to strong interdependencies between production, logistics and traffic, a decision in one of these fields has impacts on the others. Thus, a tool, which clearly illustrates the variety of impacts of a decision, is highly desirable.

In this chapter a reference model is presented to describe interdependencies and interdisciplinary impacts of measures and decisions across the disciplines, called IDM. This reference model constitutes a powerful method for structuring and analysing direct as well as indirect impacts of decisions. The IDM can serve as a decision support tool since it not only structures interdisciplinary decision variables in one comprehensive framework, but can also illustrate consequences of decisions with comprehensible decision trees. However, its flexible design allows for a high level of detail, if needed. The IDM's applicability was shown by briefly outlining several areas of application and describing some influences on production and logistics resulting from the introduction of HGV tolls originating from a decision in the traffic system.

By using the developed reference model, experts from industry, authorities and science are supported to identify the impacts of decisions on interrelated decision-makers from other disciplines. Regarding our example, this is a particular opportunity for road authorities, since they usually have no access to such information when preparing traffic-related measures. But also for decisions in production and logistics, e.g. on production schemes, supply chains or storage concepts, it is well applicable to identify the consequences for actors in traffic and transport.

Thus, for the future, the IDM for production, logistics and traffic provides a good basis for an IT-based decision support tool for planning applications in production, logistics and traffic (e.g. estimate the impacts of a truck ban on processes on production and logistics). In addition, the IDM will facilitate further research in the areas of network analysis and checklist-supported decision-making. The IDM will also provide a solid foundation for developing adequate methods for interdisciplinary decision support.

References

Anand, N., Yang, M., van Duin, J. H. R., & Tavasszy, L. (2012). GenCLOn: An ontology for city logistics. *Expert Systems with Applications, 39*(15), 11944–11960.

Anderson, S., Allen, J., & Browne, M. (2005). Urban logistics—How can it meet policy makers' sustainability objectives? *Journal of Transport Geography, 13*(1), 71–81.

Bahm, A., Leone, H., & Henning, G. (2011). SCONTO: A supply chain ONTOlogy that extends and formalizes the SCOR model. In American Institute of Chemical Engineers (Ed.), *Proceedings of AIChE Annual Meeting 2011* (pp. 920–922).

Bera, P., Burton-Jones, A., & Wand, Y. (2011). Guidelines for designing visual ontologies to support knowledge identification. *MIS Quarterly, 35*(4), 883–908.

Bertalanffy, L. V. (1976). *General system theory*. New York: Braziller.

Blecker, T., & Kersten, W. (2006). *Managing risks in supply chains: How to build reliable collaboration in logistics*. Berlin: Erich Schmidt Verlag.
Boltze, M., Dinter, M., & Schöttler, U. (1994). The project FRUIT—A goal-oriented approach to traffic management in Frankfurt am Main and the Rhein-Main region. *Traffic Engineering and Control, 35*(7/8), 437–444.
Broaddus, A., & Gertz, C. (2008). Tolling heavy goods vehicles. *Transportation Research Record, 2066*, 106–113.
Clausen, U., Iddink, U., & Neumann, L. (2008). Approaches for modelling commercial freight traffic regarding logistics aspects. In *3rd International Symposium on Transport Simulation*. Surfers Paradise.
Condeço-Melhorado, A., Gutiérrez, J., & García-Palomares, J. C. (2011). Spatial impacts of road pricing: Accessibility, regional spillovers and territorial cohesion. *Transportation Research Part A, 45*(3), 185–203.
Conway, A., & Walton, C. M. (2009). Policy options for truck user charging. *Transportation Research Record, 2115*, 75–83.
Craighead, C., Blackhurst, J., Rungtusanatham, M., & Handfield, R. (2007). The severity of supply chain disruptions. Design characteristics and mitigation capabilities. *Decision Science, 38*(1), 131–156.
Doll, C., & Schaffer, A. (2007). Economic impact of the introduction of the German HGV toll system. *Transport Policy, 14*(1), 49–58.
Einbock, M. (2006). Effects of the Austrian road toll system on companies. *International Journal of Physical Distribution and Logistics Management, 36*(2), 153–169.
European Parliament. (2008). *Pricing systems for road freight transport in EU member states and Switzerland*. Brussels: European Parliament.
Friedrich, H. (2010). *Simulation of logistics in food retailing for freight transportation analysis*. Dissertation, Karlsruhe Institut für Technologie.
Gustafsson, I., Cardebring, P. W., & Fiedler, R. (2006). *Road user charging for heavy goods vehicles—An overview of regional impact*. Hamburg: East West Transport Corridor.
Hensher, D. A., & Puckett, S. (2008). Assessing the influence of distance-based charges on freight transporters. *Transport Reviews, 28*(1), 1–19.
Krause, D. (2005). *Luhmann-Lexikon: Eine Einführung in das Gesamtwerk von Niklas Luhmann*. Stuttgart: UTB.
Link, H. (2008). Acceptability of the German charging scheme for heavy goods vehicles: Empirical evidence from a freight company survey. *Transport Reviews, 28*(2), 141–158.
Luhmann, N. (2011). *Organisation und Entscheidung*. Wiesbaden: Verlag für Sozialwissenschaften.
Maier, R., Hädrich, T., & Peinl, R. (2005). *Enterprise knowledge infrastructures*. Berlin: Springer.
Manheim, M. (1979). *Fundamentals of transportation systems analysis*. Cambridge, MA: The MIT Press.
March, J. G. (1965). *Handbook of organizations*. Chicago, IL: Rand McNally.
Martínez Sánchez, A., & Pérez Pérez, M. (2001). Lean indicators and manufacturing strategies. *International Journal of Operations and Production Management, 21*(11), 1433–1452.
McKinnon, A. (2006). A review of European truck tolling schemes and assessment of their possible impact on logistic systems. *International Journal of Logistics: Research and Applications, 9*(3), 191–205.
Miranda, H. F., & Rodrigues da Silva, A. N. (2012). Benchmarking sustainable urban mobility: The case of Curitiba, Brazil. *Transport Policy, 21*(5), 141–151.
Musso, A., & Rothengatter, W. (2013). Internalisation of external costs of transport—A target driven approach with a focus on climate change. *Transport Policy, 22*(9), 303–314.
Pfohl, H.-C., Ehrenhöfer, M., & Berbner, U. (2012, August 15–17). Fundamental problematic issues in interdisciplinary decision-making: An integrated view on production, logistics and traffic and transport. In *Proceedings of the 9th International Meeting on Logistics Research*. Montreal.
Pfohl, H.-C., Zuber, C., Berbner, U., & Ehrenhöfer, M. (2013). Dynamic and seamless integration of production, logistics and traffic/transport—Barriers to interdisciplinary decision-making.

In World Conference on Transport Research Society (Ed.), *Selected Proceedings of the 13th World Conference on Transport Research Society*.
Poluha, R. G. (2007). *Application of the SCOR model in supply chain management*. London: Cambria Press.
Porter, M. E. (1985). *Competitive advantage: Creating and sustaining superior performance*. New York: Free Press.
Roggisch, N., & Wyssuek, B. (2002). Systeme und Modelle. In H. Krallmann, H. Frank, & N. Gronau (Eds.), *Systemanalyse im Unternehmen: Vorgehensmodelle, Modellierungsverfahren und Gestaltungsoptionen* (pp. 21–46). München: Oldenbourg Wissenschaftsverlag.
Roorda, M. J., Cavalcante, R., McCabe, S., & Kwan, H. (2010). A conceptual framework for agent-based modelling of logistics services. *Transportation Research Part E: Logistics and Transportation Review, 46*(1), 18–31.
Rühl, F., & Boltze, M. (2016). Freight transport demand management—Influencing the freight transport demand within traffic management. In E. Abele, M. Boltze, & H.-C. Pfohl (Eds.), *Dynamic and seamless integration of production, logistics and traffic. Fundamentals of interdisciplinary decision support* (pp. XXX). Berlin: Springer.
Rühl, F., Freudenreich, T., Berbner, U., Ottemöller, O., Friedrich, H., & Boltze, M. (2013). Production, logistics, and traffic: A systematic approach to understand interactions. In World Conference on Transport Research Society (Ed.), *Selected Proceedings of the 13th World Conference on Transport Research Society*.
Sanders, N. R., Zacharia, Z. G., & Fugate, B. S. (2013). The interdisciplinary future of supply chain management research: Interdisciplinary future of SCM research. *Decision Sciences, 44*(3), 413–429.
Stathopoulos, A., Valeri, E., & Marcucci, E. (2012). Stakeholder reactions to urban freight policy innovation. *Journal of Transport Geography, 22*, 34–45.
Supply-Chain Council. (2010). *Supply chain operations reference model*. Version 10.0.
Thom, A. (2008). *Entwicklung eines Gestaltungsmodells zum Management von Risiken in Produktionsnetzwerken: ein Beitrag zum Risikomanagement in der Logistik*. Berlin: Univ.-Verl. der TU Berlin.
Vasallo, J. M., & López, E. (2010). Using input-output tables to estimate the effect of charging heavy goods vehicles on CPI. *Journal of Transport Economics and Policy, 44*(3), 317–329.
Vidaillet, B. (2008). When "decision outcomes" are not the outcomes of decisions. In G. Hodgkinson & W. Starbuck (Eds.), *The Oxford handbook of organizational decision making* (pp. 418–437). Oxford: Oxford University Press.
Wang, C., & Masini, A. (2009). *The sand cone model revisited: The impact of service flexibility on quality, delivery, and cost*. Working paper, London Business School.
Zuber, C., Pfohl, H.-C., & Berbner, U. (2015). Integrating domains in supply chains: Development of requirements for interdisciplinary decision support based on an integrative framework of production, logistics, and traffic. In European Decision Sciences Institute, J. Stentoft, A. Paulraj, & G. Vastag (Eds.), *Research in the decision sciences for innovations in global supply chain networks: Best papers from the 2014 Annual Conference* (pp. 233–254). Upper Saddle River, NJ: Pearson FT Press.

Part II
Interdisciplinarity from a Domain Specific Perspective of Production, Logistics and Traffic

Flexible Design of Lean Production Systems in Response to Fluctuations Due to Logistics and Traffic

Stefan Seifermann, Jörg Böllhoff, Siri Adolph, Eberhard Abele, and Joachim Metternich

Introduction

A value creating supply chain of physical products, reaching from various suppliers to different customers, involves several actors in the fields of production, logistics and traffic. This diversity of stakeholders naturally results in a number of interfaces between the parties. An interface is always a possible point of discontinuity. In most cases, interfaces need to be actively managed in order to minimize disturbances, and decisions on both sides of the interface have to be aligned. However, the vast majority of decisions across production, logistics and traffic are taken independently within the discipline, deliberately or undeliberately ignoring effects on other stakeholders. This course of action leads to unplanned disruptions in the chain.

Amongst others, industrial production plants suffer severely from disturbances. Especially in unbuffered production systems that are set up according to the principles of lean production, stability is a crucial prerequisite. Due to very low inventory levels, any unplanned disturbance in a lean production environment directly affects the actual production order, ultimately resulting in a production stop and reduced performance levels of so-called Key Performance Indicators (KPIs) like costs, quality and delivery.

One approach for at least partially absorbing the effects of unplanned disturbances on production is raising the overall flexibility levels within production itself. The focus of this chapter is thus set to how flexibility can be leveraged in production systems to deal with disturbances resulting from fluctuations in logistics and traffic. In order to get an impression of the flexibility requirements, the considerations start

S. Seifermann (✉) • J. Böllhoff • S. Adolph • E. Abele • J. Metternich
Institute of Production Management, Technology and Machine Tools, Technische Universität Darmstadt, Otto-Berndt-Str. 2, 64287 Darmstadt, Germany
e-mail: seifermann@ptw.tu-darmstadt.de; boellhoff@ptw.tu-darmstadt.de; adolph@ptw.tu-darmstadt.de; abele@ptw.tu-darmstadt.de; metternich@ptw.tu-darmstadt.de

E. Abele et al. (eds.), *Dynamic and Seamless Integration of Production, Logistics and Traffic*, DOI 10.1007/978-3-319-41097-5_4

with mapping the supply chain and describing the nature and severity of potential fluctuations using the fault-tree analysis method. Currently, most production systems are decoupled from logistics and traffic by excessive inventory buffers. Consequently, the effects of fluctuations on production from these disciplines are absorbed at the costs of overcapacity in people, inventory, processes and machinery. The concept of lean production identifies and addresses these overcapacities as "waste" and is therefore taken as a basis for further improvement discussions in the chapter.

After having developed the flexibility needs, a closer look at the term "flexibility" itself in manufacturing contexts as well as at the different types of flexibility will be taken. Volume and mix flexibility are identified as the types most relevant for industrial companies. Thus, different models for measuring volume and mix flexibility in the literature will be reviewed. The chapter then turns to the actions for increasing flexibility in production systems and presents selected technical, organizational and human approaches. The focus is from then on set to lean production systems and their special conditions and prerequisites. First, selected approaches for addressing volume and mix flexibility in lean assembly systems will be discussed. Afterwards, the innovative concept of Cellular Manufacturing as an approach to transfer the ideas from lean assembly to lean machining will be introduced. In Cellular Manufacturing lines, a complex machining task, usually executed on one machine tool in complete machining, is split into several fragments being distributed to a number of smaller, specialized and cheaper machine tools adequate for the operation (right-sized equipment). Volume and mix flexibility advantages due to the Cellular Manufacturing concept will be presented. As any flexible concept is only viable in industrial practice if it is also economically advantageous, the chapter concludes with a cost comparison of Cellular Manufacturing versus complete machining.

Flexibility Requirements in Production due to Logistics and Traffic

In order to identify the appropriate flexibility actions in production as responses to fluctuations in logistics and traffic, it is first necessary to get an overview on potential disruptions along the supply chain. Figure 1 exemplarily shows an extract of a typical value creating supply chain of physical products, comprising production, logistics and traffic. Several actors like producing plants, warehouses, various transportation routes, etc., are visualized.

Interdisciplinary disruptions are either caused by occurrences or behaviours within one discipline, which are unexpected and unplanned in the other one, or by not adequately managed interfaces. While the responsibilities for disruptions with implications restricted to the generating discipline itself are relatively easy to

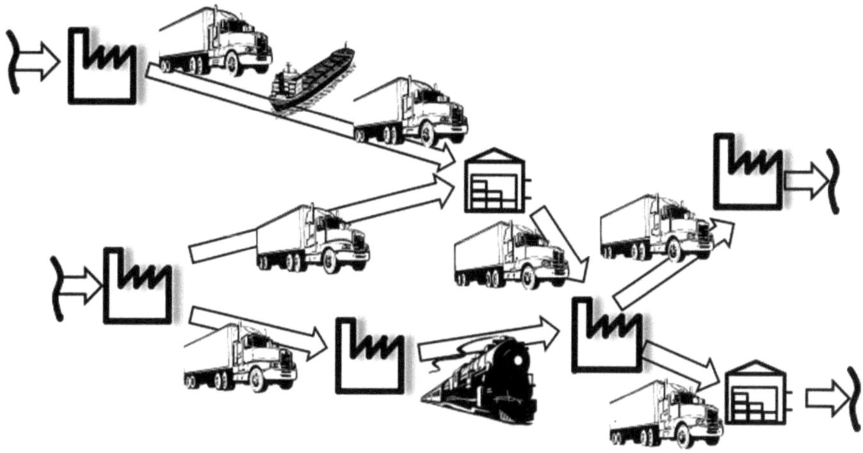

Fig. 1 Exemplary excerpt of a supply chain along production, logistics and traffic

address, disruptions with consequences crossing the interfaces to other disciplines are more difficult to handle: Responsibilities are often not clear, and the power to initiate changes in other disciplines than one's own is very restricted, especially in dealing with the publicly managed traffic sector. Therefore, the interfaces between production, logistics and traffic are amongst the weakest points for disturbances in the supply chain.

For the production discipline, two types of interfaces to logistics and traffic have to be distinguished:

- For physical materials, the dock doors of receiving and of shipping areas inside the production plants are the points in the supply chain, where goods are transferred from logistics and traffic to production.
- For information being exchanged between the disciplines, production and transportation planning departments inside the production organizations are responsible.

Events in logistics and traffic with disturbing implications on production across the interfaces mentioned above are different in time of notice, severity and nature. Figure 2 gives a rough overview on interrelations between time of notice, severity and flexibility requirements.

In order to distinguish and to classify the various natures of disturbances in logistics and traffic qualitatively, Fig. 3 shows exemplary results of a fault-tree analysis conducted on the topic. The fault-tree analysis method was developed by the Bell Telephone Laboratories in 1961 for systematically detecting potential failures of technical systems and the basic causes of these failures (Thums 2004). In a top-down approach, an abstract event is fragmented step by step into all potential root causes on different levels of aggregation (Bertsche and Lechner 2004).

On the first level, the fault-tree distinguishes transport and storage of goods in warehouses. For all transports, the primary goal is bridging the gap between one

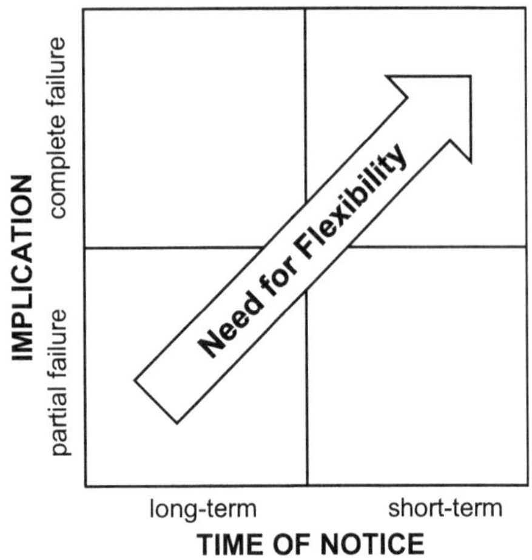

Fig. 2 Systematization of required flexibility, depending on time of notice and the implications

location and another (Martin 2014; Aberle 2003). External transport can be further broken down into different modes of transport (Aberle 2003); internal transport can be split into organizational and modal categories (Martin 2014). Especially for the means of conveyance exists a multitude of further classification approaches (Jünemann and Schmidt 2000). The storage of goods in warehouses can serve different purposes: On the one hand, goods are stored in warehouses in order to accumulate inventory long-term or for temporary transition. On the other hand, warehouses are used to rearrange and distribute material flows. In both types of warehouses, the basic processes are identical, but of different importance (Pfohl 2010). Thus, loading and unloading goods, picking and storage itself can be identified (Gudehus 2010). The basic level of the fault tree exemplarily shows potential individual events in logistics and transport that can lead to disturbances in production.

For industrial companies, there are various possible ways of dealing with disturbances coming from beyond the interfaces to logistics and traffic. However, actual approaches in the industry—even in just-in-time environments—currently focus mainly on decoupling the different actors along the supply chain at the interfaces, namely via the installation of smaller or larger warehouses or other buffers before and after production. Subsequently, the need for alignment of processes on both sides of the interface tends to be low. This results in excess overcapacity, inventory, processes and machinery for all stakeholders.

The majority of improvements due to a flexibility increase in production will be generated at the inbound interface between production and logistics/traffic. There, the effects of disturbances in production coming from logistics and traffic are much more significant than on the downstream outbound interface. All inbound disturbances can easily lead even to production stops, while outbound disturbances caused by logistics or traffic will at worst result in a pile-up of goods at the outbound warehouse.

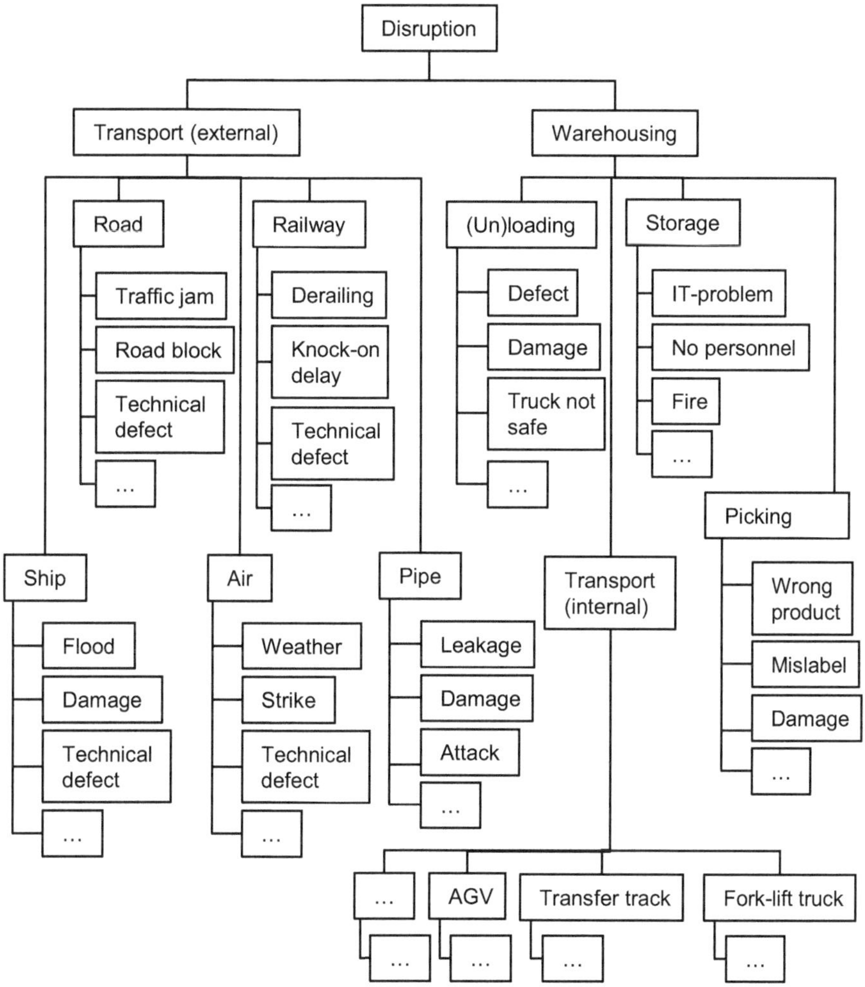

Fig. 3 Qualitative fault-tree analysis of disruptions in logistics and traffic leading to potential issues in production (adapted from Schulze 2012)

In recent years, the concept of Lean Production has gained more and more importance in industrial production. The dictum of this concept is a minimization of lead times achieved by reducing or even eliminating any waste along the whole supply chain (Ohno 1993). Thus, also overcapacities at the interfaces of production, logistics and traffic are addressed. One approach within Lean Production is an increase in flexibility in order to absorb the effects of fluctuations entering the producing system. The following explanations in this chapter will examine the state of the art of flexibility in production and how it can be used to improve the whole supply chain.

Flexibility in Production: State of the Art

The Term "Flexibility" in Manufacturing Contexts

In manufacturing contexts, the term "flexibility" is used in several, partially even contradictory ways. A number of definitions are at hand (see for example Kaluza and Blecker 2005; Van Hop 2004; Zelenovic 1982). Depending on the focus of the individual research, these definitions emphasize slightly different aspects of flexibility. Following the comprehensive definitions of De Toni and Tonchia (1998) and Nyhuis et al. (2008), manufacturing flexibility shall be understood as a manufacturing system's predefined capacity to respond to expected changes in requirements within a certain band width by moving quickly and with little penalty in time, effort, cost or performance from one state to another. With this understanding, flexibility can be clearly distinguished from reconfigurability: While flexibility merely includes potential reactions within certain limits to changes that can be foreseen at the time of planning, reconfigurability also includes possible reactions to changes in the environment that are unforeseen and push the system beyond the flexibility limits (Nyhuis et al. 2008).

Manufacturing flexibility can be further classified into several different types. Referring to Browne et al. (1984) and Sethi and Sethi (1990), ElMaraghy (2005) distinguishes the following partially interrelated classes:

- Machine flexibility: the same machine's ability to execute various operations without any changeover effort.
- Material handling flexibility: refers to the material handling system's capacity to be used for individual part types and alternative routes.
- Operation Flexibility: ability of one part to be produced in different alternative ways.
- Process Flexibility/Mix Flexibility: the same machine's ability to produce different predetermined part types without major changeover effort.
- Product Flexibility: ease of changing the product mix by introducing new or modified parts.
- Routing Flexibility: ability of a system to produce a part along different successive routes, i.e. on different machines, with different operations, or in different sequences.
- Volume Flexibility: ability to operate profitably at varying demands.
- Expansion Flexibility (consistent with reconfigurability): Ease of physically changing the capacity and/or capability boundaries of a production system.
- Control Program Flexibility: ability of a manufacturing system to operate unattended by a person.
- Production Flexibility: range of part types that can be produced by the system without major investments.

Several of the types of manufacturing flexibility mentioned are of a more theoretical nature, and the importance for the industrial practice greatly varies.

Approaching the topic from the customer side, Suarez and Cusumano (1996) define four "first-order" flexibility types: mix flexibility, volume flexibility, new-product flexibility and delivery-time flexibility. However, delivery-time flexibility is neither mentioned as a "classic" flexibility type, nor even further explained by Suarez et al. Rather it is a resultant of the other three first-order types. New-product flexibility is greatly influenced by product development processes, and thus not part of the day-to-day operation. It will not be further considered. The research for a flexible design of lean production systems in response to fluctuations due to logistics and traffic will concentrate on volume flexibility and mix flexibility. This emphasis is also in line with our numerous observations of flexibility understandings in the industrial practice and with recent research on the importance of these flexibility types (Lanza et al. 2009).

Evaluation of Volume and Mix Flexibility in Manufacturing

Similar to the strongly varying perceptions of the term flexibility, differing models for measuring flexibility exist. An overview as well as a classification of the individual models can be found in Bellmann et al. (2009) or in Gupta and Goyal (1989). This chapter details on the different efforts carried out in the past for evaluating volume and mix flexibility in production. Benefits and downsides of the individual approaches are mentioned.

Selected Evaluation Models for Volume Flexibility

In discussions on "volume flexibility", there is always the reference to "profitability": Volume flexibility describes the ability of a system to operate in a certain range of varying demands under the constraint of being economically advantageous at any point of operation (Browne et al. 1984).

Approach of Parker and Wirth (1999)

Parker and Wirth define the span between the available capacity of a production system and the financial break-even line as an adequate measure for volume flexibility. The accompanying formula for n products is

$$\text{Volume flexibility} = 1 - \frac{F}{C_{\max}}\left(\prod_{i=1}^{n}\frac{a_i}{b_i}\right)^{\frac{1}{n}}$$

(with F as fixed costs, C_{max} as production system capacity, a as the required amount of capacity units to produce one product unit and b as the contribution margin of one product unit).

However, the authors ignore demand-dependent changes in variable costs (neglecting for example economies of scale or additional costs due to premium payments in extra shifts) and assume equal shares of demand for different products. In reality, this will rather not be the case.

Approach of Das (1996)

Starting from the same basic considerations as Parker and Wirth, Das defines volume flexibility as "a measure of the company's ability to continue operating economically at less than 100 % capacity". He further argues that the economic minimum in case of an intended overcapacity (100 %) will not be the break-even point (V_{BE}/C_E), as a comparable company designed for exactly that capacity V_{BE} would be able to realize much lower costs C_H. Therefore—in a profitability sense—the minimum economic production capacity V_{min}, and thus the lower end of the volume flexibility $VFLX_{CAP}$, is the point where the actual costs up-charged by the difference between C_E and C_H (as kind of opportunity costs) meet the revenues. Figure 4 illustrates the correlations.

Although these considerations are theoretically correct, there is justifiable critique for this approach, as the company still generates profit in the area between V_{BE} and V_{min}. Any real company would therefore consider the whole span from V_{BE} to the maximum capacity of 100 % as a viable range for operation.

Approach of Beamon (1999)

Beamon's approach also takes the maximum and the minimum profitable production volumes O_{max} and O_{min} into account and links these with basic statistical

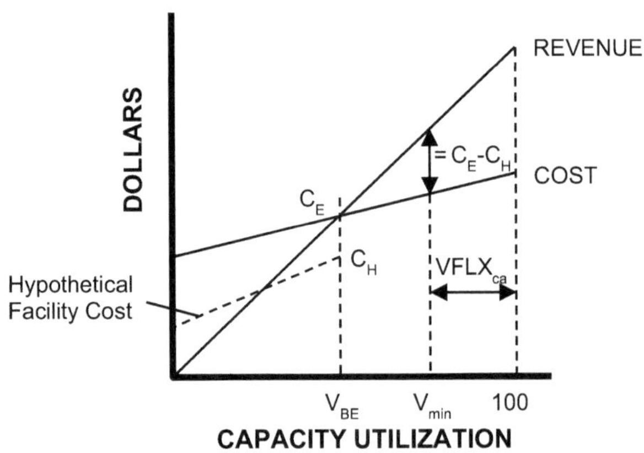

Fig. 4 Measuring volume flexibility according to Das (1996)

distribution measures. Assuming a standard distribution of demand volumes D with mean demand $\overline{D}$ and standard demand variance S_D^2, the volume flexibility can be defined as

$$\text{Volume flexibility} = \Phi\left(\frac{O_{\max} - \overline{D}}{S_D}\right) - \Phi\left(\frac{O_{\min} - \overline{D}}{S_D}\right)$$

By including the extent of the demand variation into the calculation, Beamon further details the volume flexibility measure beyond linear costs and revenues.

Selected Evaluation Models for Mix Flexibility

Mix flexibility describes the ability of a production system to produce different predefined products without major effort for changeover (Browne et al. 1984). As the products are known and designated at the time of the system's installation, mix flexibility clearly needs to be distinguished from expansion flexibility that deals with later on decisions to manufacture products on this production system.

Approach of Das (1996)

For measuring mix flexibility, Das develops a product differentiation measure for products k_1 and k_2 at machine j. He finds the differences rated on a scale from 0 to 1 in

- Processing operations: $\Delta\text{Oper}_{k1,\, k2,\, j}$
- Worker skills: $\Delta\text{Skill}_{k1,\, k2,\, j}$
- Physical nature of products: $\Delta\text{Prod}_{k1,\, k2,\, j}$
- Machine processing time: $\Delta\text{Mach}_{k1,\, k2,\, j}$

Including a weighting of the criteria with weights W_1 to W_4, the weighted aggregated difference $E_{k1,k2}$ between two products is thus

$$E_{k1,k2} = E_{k2,k1} = \frac{1}{M}\sum_{j=1}^{M}\Big(W_1\Delta\text{Oper}_{k1,k2,j} + W_2\Delta\text{Skill}_{k1,k2,j} + W_3\Delta\text{Prod}_{k1,k2,j}\Big)$$
$$+W_4\Delta\text{Mach}_{k1,k2,j}$$

The mix flexibility of the whole product portfolio with L products can then be defined as

$$\mathrm{CFLX_{cap}} = \frac{1}{L}\sum_{k_1=1}^{L}\sum \begin{matrix} k_2 = 1 \\ k_2 \neq k_1 \end{matrix}^{L} E_{k1,k2}$$

Although being theoretically sound, the major point of critique for this approach is the vague rating of the criteria from 0 to 1. Depending on the individual situation, Das holds the respective production managers and employees responsible and gives mere recommendations for concluding a rating.

Approach of Bateman et al. (1999)

Following the explanations of Chryssolouris and Lee (1992), who introduce the indicator "sensitivity to change" (STC) as reciprocal measure for a machine's flexibility, Bateman et al. further developed a mean index MSTC over all n product variants and production sequences. Leveraging the probability $P_i(P_i - 1)$ of a product i being followed by a different product and the setup duration dur of a product, they conclude

$$\mathrm{MSTC} = \sum_{i=1}^{n} P_i(P_i - 1)\mathrm{dur}_i$$

Approach of Wahab (2005)

For expressing a system's mean product mix flexibility range SPMRF, Wahab leverages a similarity index α of two different products k and l. He assumes the setup time to correlate with the number of tools being exchanged for manufacturing each type. Second, he supposes that different machines j operate with different efficiencies and products will be assigned to the most efficient machine. Consequently, he defines p_{kj} as auxiliary variable for the probability that product k is assigned to machine j. He concludes

$$\mathrm{SPMFR} = \frac{1}{m}\sum_{j=1}^{m}\sum_{k=1}^{n}\sum_{l\neq k}^{n} p_{kj}\alpha_{kl}$$

Selected Models for the Combined Evaluation of Volume and Mix Flexibility

Beyond the approaches that separately examine volume and mix flexibility, there are also some attempts for combined models. However, while volume and mix

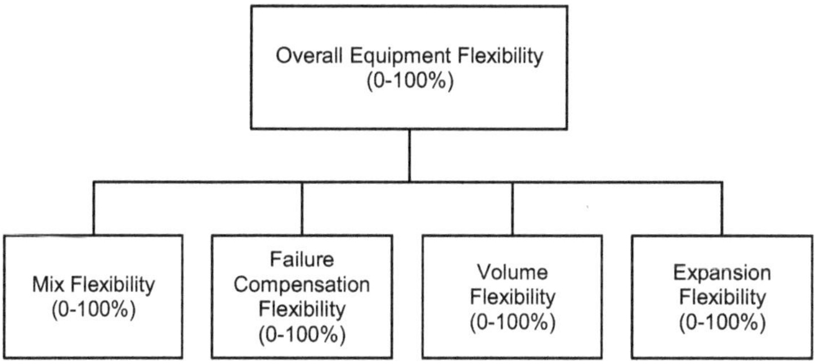

Fig. 5 Structure of the Overall Equipment Flexibility OEF and its subtypes (Abele et al. 2008a)

flexibility are the prevailing flexibility types for an industrial operation, the models mentioned partly address also the influences of other flexibility types without further distinction.

Approach of Abele et al. (2008a)

Following the well-known KPI "Overall Equipment Effectiveness" (OEE), Abele et al. propose to aggregate four different weighted measures for flexibility to one holistic indicator called "Overall Equipment Flexibility" (OEF). The individual types of flexibility addressed are shown in Fig. 5.

While the authors specifically strive for a wide use of their indicator in industrial enterprises, they unfortunately lack further information on how the individual subtypes can be calculated. Therefore, they reference other existing models in the literature without being specific.

Approach of Metternich et al. (2013a, b)

Referencing Chryssolouris' declaration that "any flexibility calculation which omits external demand [...] may be meaningless" (Chryssolouris 2006), Metternich et al. establish a combined volume and mix flexibility measure explicitly including the postulations of lean production systems. Therefore, the authors put customer orders d and production quantities p for each variant i for a certain time period t into relation and calculate the respective differences ${}_{\Delta}dp$. In an entirely flexible and lean production system, actual production would always fully meet the demand. Totalized over all n different products ${}_{\Delta}dp_t$ can be expressed as

$$_{\Delta}d^{\Delta}p_t = \sum_{i=1}^{n} \left| d_{t,i} - p_{t,i} \right|$$

Considering individual work intensities for different products that use different portions of the available capacity of a production system, the capacity demand c as weighting factor for producing one unit of product i is introduced. Thus, the volume and mix flexibility FLEX_t of a system can be calculated as

$$\mathrm{FLEX}_t = 1 - \frac{\sum_{i=1}^{n} c_i \left| d_{t,i} - p_{t,i} \right|}{\sum_{i=1}^{n} c_i * d_{t,i}}$$

with $\mathrm{FLEX}_t = 1$ expressing perfect flexibility.

Summary and Review of Existing Approaches to Measure Volume and Mix Flexibility

So far, several different approaches for measuring volume and mix flexibility have been presented. However, there seems to be mutual consensus that the existing approaches come along with minor or larger deficits. Several authors state that existing models are mainly not applicable for industrial operations as (see for example Abele et al. 2008a):

- Theoretically developed models are often simplified and neglect important aspects of real operations
- Models are complex and not always straightforward to calculate
- A multitude of input data is required
- Which partly cannot be collected at all or not with a justifiably effort

Also, existing models seem to be restricted to specific underlying conditions (Beck 2012), partially depending on the industry they were developed for (Koste et al. 2004). Again, there is major critique that customer requirements are largely neglected in most models, leading to flexibility discussions without even knowing the demand for it (Chryssolouris 2006; Beck 2012). This leads to the frequent circumstance that for flexibility decisions, decision-makers rather rely on their subjective interpretations and personal estimations of the situation than on a calculated model (Abele et al. 2008a). Beach et al. (2000) perceive the issue even more comprehensively: They account the inconsistent understandings of the underlying fundamental flexibility types for not being able to conclude a practical performance measure for flexibility.

Actions for Flexibility in Production

This chapter lists and characterizes different actions for increasing flexibility in production systems. Early findings at the London Tavistock-Institute suggest regarding a company and the production system as a "socio-technical system" (Trist and Bamforth 1951; see also Leavitt 1965). Actions for increasing flexibility can thus be technical, organizational or human centred. Subsequently, Ulich (2011) further developed the MTO approach. For achieving an ideal outcome, technical and social areas need to be matched and harmonized (Cummings and Blumberg 1987). Individual flexibility actions might not easily and isolatedly be attributed to either of the categories, as most of them also have impacts on other categories (Krüger 2004).

Technical Actions for Flexibility in Production

There are different technical measures leading to increased flexibility in production which will be briefly described below. In general, for achieving a wider profitable range in terms of volume, a variation of intensities respectively of production rates is necessary. For adapting the mix capability for manufacturing a multitude of different product types on existing machinery, the setup effort needs to be reduced. All technical actions normally require a modification of the production-related equipment (Askar 2008).

Overcapacity in Workstations/Machinery

A very widespread technical action for providing flexibility is the deliberate acquisition of machinery or workstations with excess capacity or capability. This means that the basic technical system is capable of producing faster and more variants than required for an average production (Krüger 2004). In some cases, even spare, identical equipment or workstations are held available. An interesting aspect is the combination of this parallelization of operations with different levels of automation (see below): Thus, the majority of volumes or variants can be handled via a highly automated system, while excess volume or special variants are produced with less automated equipment.

If overcapacity in workstations or equipment is integrated upfront, the adaption to volume or variant changes can be executed very fast. However, such a strategy goes along with higher investment costs and also requires adequate organizational and human flexibility levels of the production organization (Krüger 2004). Furthermore, this can also inhibit a company's continuous improvement process since the need for a higher productivity will not be as urgent.

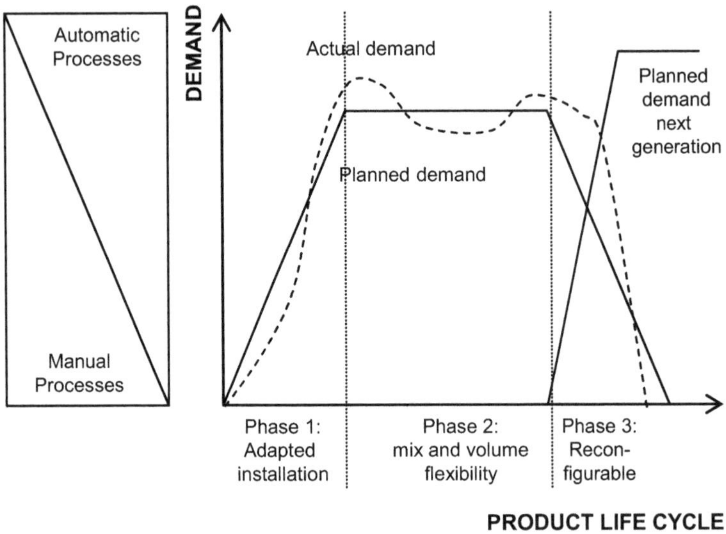

Fig. 6 Overview on incremental installation and reduction of manufacturing equipment (Lotter et al. 1995)

Incremental Installation and Reduction of Manufacturing Equipment

Aiming at a financial risk reduction in capital investment for production equipment under uncertain demands, Lotter et al. (1995) propose a gradual, adapted installation strategy. The approach shown in Fig. 6 foresees especially in ramp-up and ramp-down phases at the beginning and the end of a product life cycle manually driven processes, while during the core production phase with relatively stable, high demands, automation efforts are prevailing. Further on, the authors call for an upfront planning of manufacturing equipment with varying demand scenarios as well as for an integrated strategy to reuse machinery no longer needed in case of lower demands.

Despite preparations, measures in this area are still time-consuming and not likely to provide a fast reaction possibility to changing demands in volume or variants. Depending on the nature of the production technology and production processes, measures are rather complex to implement (Schmeißer 2012).

Variation of the Level of Automation

Strongly correlated with the gradual investment strategy introduced above is the well-directed variation of the level of automation. The goal of minimizing financial risk is again the main driver. While the human worker is—and will be at least medium-term—the most flexible resource in manufacturing (Liker 2013), automated solutions are from a certain level of demand on more cost-attractive, but inflexibly specialized (Chryssolouris 2006). In the past, hybrid systems consisting of manual and automated tasks have proven to be most successful in terms of resolving the trade-off between flexibility and efficiency (Lotter et al. 1995). This combined approach leaves enough room for an adaptation of individual work content to either more or less volume or special tasks for particular product variants.

Organizational Actions for Flexibility in Production

In general, organizational measures can be characterized as independent of related production processes (Askar 2008). The split between organizational and human-centred actions is rather weak: Some measures can be classified as either organizational or human centred or mutually require actions in the other field.

In- and Outsourcing of Work Content

The approach of shifting work to and from internal or external suppliers is very common among industrial companies. Both capacity or capability can be increased or reduced and thus be adapted to the actual need (Krüger 2004). Varying the own vertical depth of production is normally associated with deviations in costs. Finding an adequate supplier and formally ensuring its qualification for delivering the products to certain company standards, however, may take time. Therefore, actions in this area are also at least mid-term focused.

Inventory Build-up and Reduction

Increasing flexibility by building and reducing inventory belongs to the category of temporal flexibility measures: raw materials are ordered or processed before there is an actual, timely demand. The demand is expected for future time periods, when the company lives on the stock built up earlier (Jack and Raturi 2002). This is especially true for seasonal businesses (Abele et al. 2008b). Next to flexibility, inventories serve enterprises with various functions, namely cost reduction by eliminating stock-out-driven costs and costs related to any distress caused by stock bottlenecks, security or delivery time reduction. Companies can accumulate inventories as raw materials, intermediate products or finished product (Stölzle et al. 2004). However, there are two major points of concern with inventories:

- Inventories come along with certain costs of annually 16–26 % of its value for bound capital, damages and losses, administration, insurance, etc. (Stölzle et al. 2004; Shazly 2011).
- Inventories might conceal issues in production and thus hinder the elimination of waste (Pfeiffer and Weiß 1994).

As described above, increasing flexibility by providing enough inventory to cover up any fluctuations in the whole supply chain along production, logistics and traffic is a quite common measure in industrial reality. Therefore, this fact should be given special attention in future research. Nevertheless, high inventories are especially needed when processes are not flexible enough to follow the market demand directly.

Focusing on Bottlenecks and Continuous Expansion of Capacity and Capability

A more general organizational procedure for an increased flexibility is the continuous identification and elimination of bottlenecks and, consequently, the improvement of the whole manufacturing system (Imai 2001). Under this roof, a number of different methods can be subsumed, for example the SMED method for analysing and improving product variant setups (see below) (Shingo 1989).

Human-Centred Actions for Flexibility in Production

Human-centred actions for increasing volume flexibility in production are changes in the type or number of employees and the distribution of work over time (Askar 2008). Increasing the level of mix flexibility is highly linked to qualification (Gerwin 1989). The following paragraphs will introduce several human-centred approaches for increasing volume and mix flexibility.

Flexible Working Time Arrangements

Flexible working time discussions deal with the length, the position and the distribution of working time, as well as with a company's freedom of action to conduct changes in either of these fields (Ehmer 2009; Faißt 1992). Schlick et al. (2010) identify 12 different models for the variation of daily or weekly working time including starting and ending time. These include several types of flexitime and part-time job agreements. Particularly flexitime measures are supported by broad statutory rules (Ehmer 2009). Flexible working time arrangements are especially advantageous for minor or medium, short-term volume adjustments of required capacity to a fluctuating demand (Jack and Raturi 2002).

Changes in Shift Models

A medium- or long-term action for leveraging human-centred volume flexibility is changes in shift models. With this measure, additional or fewer shifts can be created. However, this measure usually goes in line with changes in the number of workers (discussed below). In case of the installation of additional shifts on weekends or during nights, premiums are due (Schlick et al. 2010).

Changes in the Number of Employees

Varying the workforce is one of the most common practices for adapting to fluctuating demands. For changes in the number of employees, two fundamentally different types have to be distinguished: First, there is the creation or cut of regular jobs within the company or the move of personnel from one department to the other. Personnel affected by these changes hold regular work contracts with the company requiring the flexibility (Askar 2008). Any changes of these types of workers are generally costly, time-consuming and difficult to reverse. Therefore, this option is only approached, if there is a relatively high guarantee that the additional or reduced volume or variants are long-lasting (Jack and Raturi 2002).

Second, there is the form of temporarily leased contract workers, which is emerging especially in recent years. These personnel are formally employed by a different company than the ordering one they finally work at. Their operation at the ordering company is temporarily restricted. Thus, leased contract workers normally are leveraged for either low-qualified tasks in order to enhance a company's volume flexibility or for very specialized projects, for which the ordering company lacks specialists. Flexibility actions in this field are short term (Sende et al. 2011).

Flexible Assignment of Employees to Workstations/Multidimensional Skills

A frequent exchange of employees among different workstations or work teams, e.g. via job rotation, can massively enhance mix flexibility. Thus, employees get qualified for different tasks, leading to increased options for replacements and assignments in case of product variant changes (Gerwin 1989). For the development and classification of this flexibility, the number of tasks the operator is qualified to conduct is crucial. Hackstein (1989) distinguished three different administrational types of organizing the workforce:

- Complementary type: With minor effort, operators are further qualified for one or more well-defined additional workstations with a work content similar to their own workstation.
- Pool type: All employees have a common basic qualification plus individual special knowledge.
- Spare-man type: Spare-men are qualified to operate at all workstations within their team/department.

Dealing with Flexibility in Lean Production Systems

The origin for what is known in the Western world as "lean production systems" is the Japanese car manufacturer Toyota. Due to their steep rise and current domination of the global automotive industry as one of the biggest and most profitable companies in the sector, competition as well as other branches is closely examining Toyota's practices (Krafcyk 1988). The final goal of all activities within Toyota is the removal of any waste along the whole supply chain with the intent to reduce the time between the receipt of a customer order until the delivery and payment of the final product (Ohno 1993). This approach ultimately results in waste-free, synchronized production systems with minimal buffers in inventory and spare capacity. Despite the required robustness of underlying basic processes, lean production systems thus are highly fragile in terms of unplanned occurrences, e.g. due to fluctuations in logistics or traffic. As a consequence, lean production systems aim at getting unplanned variations under control, if possible, or try to adapt themselves quickly to fluctuations, if they are uncontrollable (Takeda 2009). With today's increasing number of variants and rising volatility, the required flexibility in such a lean system gains importance. In the past, flexibility has been approached differently in lean assembly and in lean machining systems. The approaches will be described in the following chapters.

Flexibility in Lean Assembly Systems

This sub-chapter will give an overview on how flexibility has been included in lean assembly systems in recent years. Compared to machining, assembly areas and tasks mostly provide a good environment for volume and mix flexibility. Richter (2006) see the reasons in:

- Comparatively low capital investment in equipment
- Higher mix flexibility of the equipment itself
- Relatively low setup times and
- Output volume often directly depending on the number of operators

One basic principle of lean production systems is the alignment of all tasks to a measure quantifying the customer demand per variant: the takt time. The takt time is calculated via (Erlach 2010)

$$\text{Takt time} = \frac{\text{Available net operating time per period}}{\text{Customer demand per period}}$$

Any additional flexibility requirements for the operation need to be included into the considerations of the takt time, either by decoupling the customer demand from the actual production via inventories or delivery delays or by flexible adaptation of the operation (Erlach 2010). The first reaction, also known as "levelling by volume" and "levelling by product" (Matzka et al. 2009), is helpful for generating stability in production, but somewhat contradictory to the principle of reducing inventory as well as avoiding flexibility (Hüttmeir et al. 2009). Levelling will therefore not be further discussed in this chapter; the focus is set on volume and mix flexibility in lean assembly.

Volume Flexibility in Lean Assembly Systems

Volume flexibility in lean assembly systems can be achieved by rigorously adapting the output of an assembly to the actual demand. In practice, this will be enabled by providing different operational setups for various expected demand scenarios. The setups have been predefined and tested before the actual need. Thus, changes from one operational scenario to the other are possible in a smooth and flawless way, ensuring flexibility simultaneously with a constantly high quality and productivity (Drew et al. 2005; Abele et al. 2008b).

One key prerequisite for preparing the different operational scenarios is distributing the work content of all assembly tasks for the product evenly over all workstations available. In doing so, the takt time calculated from the varying demands of the different scenarios represents the maximum time allowed for operating each workstation (Lane 2007).

Figure 7 shows three different demand scenarios and the flexible reaction with predefined operational setups, including corresponding takt diagrams.

Rigorous standardization and qualification strongly support the flexible setup created by different operational scenarios. Standards ensure the detailed specification and the robust repeated execution of efficient and safe operational procedures, independent from different operators. They are also fundamental for the training of operators. For a successful utilization, all personnel involved needs to be qualified in the application of the standardized work procedures up front. They are also asked to continuously improve the standards set (Drew et al. 2005). Having established the human-centred actions for flexibility which have been described above is another key factor in this flexible production system.

Beyond a lean and waste-free value chain perfectly aligned to one demand level, Drew et al. (2005) regard the adaptability to changing requirements and fluctuations mentioned above as the top challenge to achieve true leanness.

Mix Flexibility in Lean Assembly Systems

Addressing mix flexibility in lean assembly systems is equivalent to systematically reducing the changeover time necessary to produce a different product variant than the current one. Next to severe differences in work content, changeover times are to date the major obstacle inhibiting a flexible, random product sequence in a lean and balanced assembly environment. The longer a changeover takes, the more the

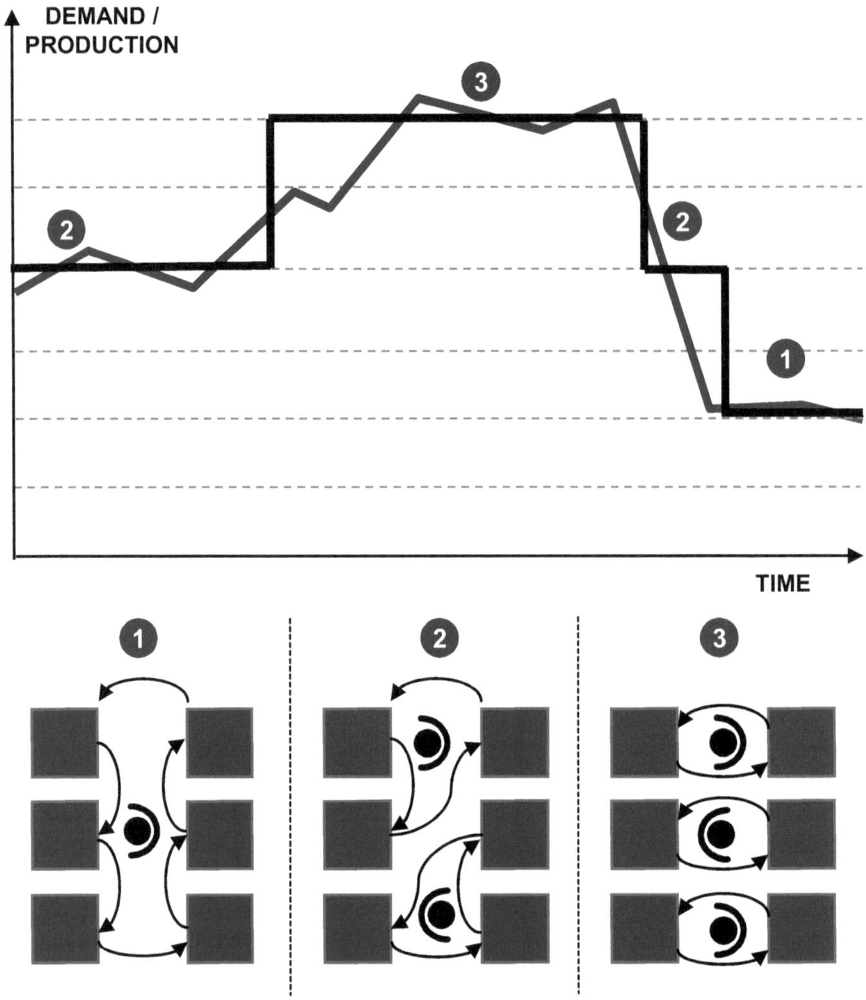

Fig. 7 Volume flexibility due to predefined operational setups for varying demands (adapted from Abele et al. 2008b; Spengler et al. 2005)

Production orders are accumulated, until a lot size large enough for an economic production run is achieved. Thus, eliminating or at least minimizing changeover efforts are crucial to produce closely to the customer demand.

Shingo (1989) proposes the method "Single Minute Exchange of Dies" (SMED) as an approach for identifying and reducing the work content during machine stop due to a type change (internal setup operations). In a first step, internal setup operations are distinguished from external ones. Afterwards, internal setup operations are converted into external activities. Again, standardization is the final step to obtain long-term, stable results.

Mix and Volume Flexible Chaku-Chaku Cells in Lean Assembly Systems

Chaku-Chaku cells are a special type of manufacturing cells. With assembly stations arranged along the sequence of assembly, all products pass through the cell in a one-piece flow. Ideally one product is assembled at each station at one point in time and each product following the sequence without passing other products. This limits the work in process to the number of stations. Generally, all assembly stations are either equipped with well-adapted, right-sized assembly automation, or—in case of mix flexibility or low investment requirements—manually operated. Especially the product transport between the individual stations is done by an operator. The work content of each station is balanced to the takt time. A U-shaped arrangement of the stations enables short transport routes. “Chaku” is the Japanese word for “loading”, implying that the operator walks from station to station and just loads and starts the next assembly process (Spengler et al. 2005; Sekine 1995; Takeda 1996).

Two operating modes of Chaku-Chaku cells exist: (Takeda 1996)

- A caravanning mode, in which the same operator passes all workstations with the same product.
- A distributed mode, in which one operator is responsible for a certain number of stations that he passes, while other stations are operated by a different person. The product is transferred to the other operator while passing the cell.

One of the predominant principles of Chaku-Chaku cells is that the machine operation is faster than the human operating time. Thus, the assembly machine is waiting for the operator, who is the bottleneck of the system. With this being true, the volume output of the cell can be varied linearly with more or less operators. Chaku-Chaku cells are volume and mix flexible and suitable for a customer-oriented assembly (Spengler et al. 2005).

Flexibility in Lean Machining Systems

While the methods and principles of lean production systems are accepted and quite widespread in assembly areas, there seems to be a significant lack of their application in machining (Abele et al. 2010). At first sight, this backlog is surprising: the advantages of lean production are hardly questionable, and at least from a theoretic point of view, all lean production methods are also applicable to machining areas. The same statement is true for the utilization of the lean flexibility concepts of takting and balancing, scenario building and setup time reduction described in the previous chapter for lean assembly systems. Although theoretically applicable, the concepts are either observed rarely in the industrial machining practice, or if at all, only with a minor scope, rigour and success.

A very good example is the SMED method: This method has originally been invented for resolving problems in the changeover of machines and has proven to provide excellent results in terms of changeover time reduction and mix flexibility (Shingo 1989). However, although being used frequently, the results lag behind those achieved in assembly, e.g. in the final car assembly, where a multitude of variants can be assembled in a takted and balanced sequence. In contrast, changeover time and its reduction in machining are often reduced to the capability of the technical equipment, organizational questions are neglected.

Abele et al. (2010) see the reasons for the machining area lagging behind in the application of lean production methods mainly in the considerably higher complexity of machining versus assembly. The advanced technical equipment—machine tools, manufacturing technology, tools, measurement technology and the like—requires an extensive knowledge and competence. Waste is harder to identify and not straightforward to eliminate.

The current state in machining operation is corresponding:

- Lot sizes in machining remain relatively high. A one-piece flow production, which is state of the art in other manufacturing areas like assembly, is observed close to never, unless in single-item production, or in inflexible automatic transfer lines.
- Changeovers from one product variant to the other are executed regularly, but not frequently and are time-consuming. Machine tool manufacturers offer extensive and expensive specialized equipment for setups parallel to the primary processing time.
- Operators are often responsible exclusively for one machine tool. In line with the two observations above, the runtime of the—mainly automatic—machine tool is quite long, leaving the operator with spare time that is currently filled with additional non-core tasks like deburring of parts, etc.
- The machine tools and the automation equipment used are quite capital intensive. In order to economically deal with the depreciation costs accompanying the high initial investments, companies aim at a full use of the machinery's capacity. Demand reductions and the resulting utilization below capacity can hardly be absorbed financially. Any demand increases on the other hand need to be delayed or outsourced to a different machine or even a different company (Abele et al. 2012).

In line with the requirements of this market, machine tool manufacturers currently pursue two different strategies. The first strategy focuses on specialized, fully automated high-performance machines that include a variety of different functions and even different manufacturing technologies. The machines built according to this strategy are usually of big dimensions and thought to be operated in stand-alone setups for a complete machining of the products (Abele et al. 2012, 2013b; Jalizi et al. 2009). The primary focus of the machine manufacturer is on the technical feasibility, not on the operability. Thus, the resulting machines are expensive, and mix flexibility is limited as the machines are normally complicated to set up.

The second strategy is purely volume driven. Machine tools are built for a mass production of the same or very similar products. Again, full automation, potentially up to automatic transfer lines, is essential. These machines are also expensive and severely inflexible.

The machines resulting from both strategies and the current state of the art in machining operation described above are favourable for high volumes, few and little changes and a limited number of product variants. In terms of volume and mix flexibility, there are clear deficits of current machining concepts (Abele et al. 2011, 2012). Due to the trend of individualization, an ongoing increase of product variants, the customer's expectation of a further reduction of delivery times, the current concepts presented for machining and the resulting inflexibility of machining environments signals a strong need for action (Abele et al. 2010).

Cellular Manufacturing as an Approach for Lean Machining Systems

With the goal of leveraging the flexibility advantages of lean assembly systems described above also for machining areas, the concept of Cellular Manufacturing has been developed. The approach is a transfer of the basic principles underlying Chaku-Chaku-Cells from assembly to machining. Cellular Manufacturing builds upon the ideas of Group Technology that was originally introduced by Mitrofanov (1966) and Burbidge (1975). Their basic idea for productivity gains versus other operational concepts is grouping similar products to a family. Subsequently assuming similar production processes and sequences of these parts within one family, the equipment and workstations for the manufacture of the family are also grouped together in one location, the Cellular Manufacturing line (Irani 1999).

There are several implications for the actual operation of a Cellular Manufacturing line. Contrary to the complete machining approach, a key prerequisite of Cellular Manufacturing is the split of one complex machining task into different parts that can be machined in sequence on different machine tools. The runtimes of these parts need to be aligned to the takt time (Abele et al. 2011). In order to be able to distribute the work content evenly and thus achieving a balance over all workstations, the lengths of the individual runtimes should further be as close as possible to each other. Manual operation tasks are potentially balanced as well (Abele et al. 2012). Figure 8 shows a principal example. Line balancing and its optimization, however, is not an easy task. A multitude of scientific approaches in the field of Operations Research are available. They are often bound to very specific constraints (Halubek 2012; Boysen et al. 2007).

The machine tools in one Cellular Manufacturing line are set up according to the sequence of the operation. The goal is to achieve a single-directional one-piece material flow. Depending on the individual characteristics of the product variants machined in the line, a "jumping", leaving out of single workstations, or a "backtracking" against the intended material flow might be necessary (Milling and Thun 2005). The machines should be arranged in a U, V, L or O shape for

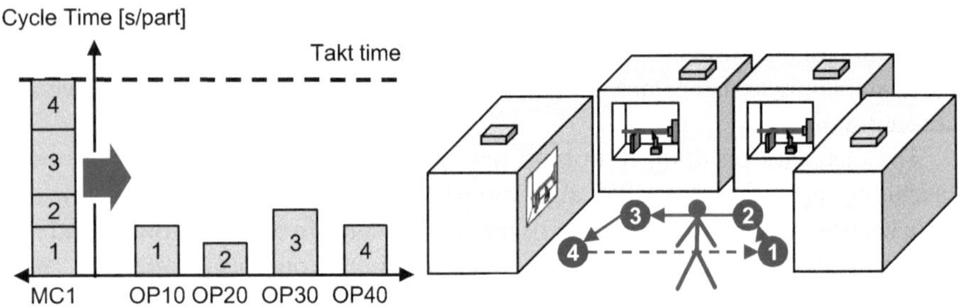

Fig. 8 Split of a complex machining task into sequential fragments and the distribution to the Cellular Manufacturing line (adapted from Abele et al. 2012)

reduced transport routes and enforced communication within the line (Shingo 1989).

Human factors play an important role in the Cellular Manufacturing concept. Same as in Chaku-Chaku assembly cells, at least the transport of the product and the operation of the equipment are executed manually (Abele et al. 2013b). Milling and Thun (2005) see the integration of human operators in Cellular Manufacturing as the most crucial success factor for gaining flexibility. They even discourage from a wider use of automation, as automation and manual operation are difficult to align. However, there is significant critique that operators in Cellular Manufacturing concepts are stressed by continuous production pressure while, simultaneously, the work itself can be quite monotonous (Schlick et al. 2010). This can be antagonized by including additional tasks like maintenance, logistics or quality assurance, etc., into the regular work (Metternich et al. 2013a, b).

Despite the postulation that all necessary processes for manufacturing a product family should be integrated into the Cellular Manufacturing line to achieve independence (Milling and Thun 2005), the split of the originally complex machining task into several fragments results in reduced requirements of the individual fragments regarding the capability of the assigned production equipment. The complexity of the individual production machine can therefore be significantly lower, as the processes are more restricted and specialized. Thus, a number of smaller dedicated machine tools with basic functions can be utilized (Abele et al. 2012). This is also described as "right-sized equipment" approach (Zelinski 2006). The capital investment for this kind of basic machine tools is comparably low. Therefore, the resulting lower depreciation costs lead to a relatively low percentage of fixed costs from total manufacturing costs. An economic operation, even in times of lower demand, is still given for a short period of time (Abele et al. 2012).

Volume and Mix Flexibility in Machining with Cellular Manufacturing

Cellular Manufacturing lines represent a good fundament in machining for dealing with fluctuations and changes resulting from a volatile environment. Especially for variations in customer demands, Cellular Manufacturing concepts show superior results. There are two main approaches for providing volume flexibility (Metternich et al. 2013a, b):

- Adapting the number of operators working in the Cellular Manufacturing line
- Adapting the number of machine tools used in the Cellular Manufacturing line

At increasing demands, the first approach can be leveraged with adding operators, as long as the manual processes are the bottleneck ("pacemaker") of the system. The upper limit of demand increases addressed with adapting the number of operators is in reaching the maximum capacity of the machinery in a 24/7 operation. At decreasing demands, operators can be drawn from the line. The lower limit for this action is the economic level at which the fixed costs from underutilized machines are still bearable. As stated above, depreciation can be kept low by using right-sized equipment. Low capital investments are therefore essential for volume flexibility in this concept (Metternich et al. 2013a, b).

The second approach, adapting the number of machines to the changing demand, requires a revised distribution of machining tasks to the different machines each time the number of machines is actually changed. This re-balancing can consume time and effort. A high degree of freedom for separating the work contents of the machining tasks is helpful.

In case of demand decreases, machines can be eliminated from the line. Thus, the capacity offered also decreased; the new takt time to which the work content is balanced increases. In times of rising demands, machines can easily be added. Due to the reasonable costs of the right-sized equipment, the purchase of an additional machine should be manageable (Metternich et al. 2013a, b; Abele et al. 2012). Both scenarios are further simplified, if a pool of machine tools is available, with which several Cellular Manufacturing lines can exchange machines. Again, small and lightweight machines that might be moved by a manual lift truck or by a hover cushion reduce the effort.

Exemplarily, Fig. 9 shows how complete machining and Cellular Manufacturing deal with volume increases and decreases. In the middle column, perfectly balanced systems are presented. To the left, the demand slightly decreases, leading to a higher takt time. The right column is converse. While complete machining concepts are faced with either underutilization (left) or the investment in a full scope machine (right), the Cellular Manufacturing approach can be adapted with a reasonable effort (Abele et al. 2012).

Mix flexibility in Cellular Manufacturing is enabled by a generally low level of automation and the manufacturing of similar products in one line (Milling and Thun 2005; Abele et al. 2010). Consequently a changeover of such a machining cell is narrowed to the machines in line. There is no need to changeover complex

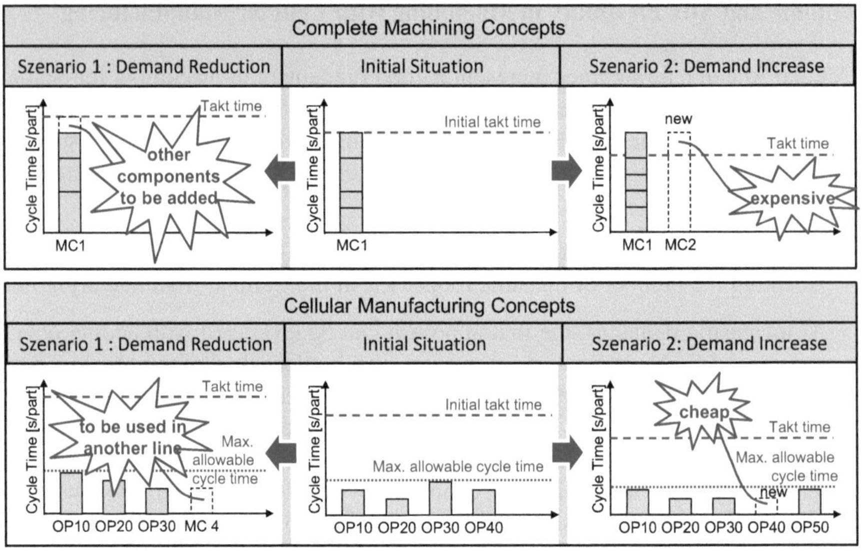

Fig. 9 Volume flexibility of complete machining and Cellular Manufacturing concepts (adapted from Abele et al. 2012)

automation equipment. Furthermore, not only the use of flexible human workforce supports the mix flexibility of the cell. Due to having more than one machine, there are also several options for efficiently implementing variant changes. First of all, less tools are needed in the individual machines because the work content of the complete machining task is split into fractions. Second, the implementation of one-piece flow enables using smaller clamping devices because only one unit needs to be fixed at a time. This provides the possibility to leave clamping devices for several variants on the machines and not take them out during changeover. Finally, setups on basic machines are easier to carry out and can even be executed in parallel on all machines.

Economic Aspects of Cellular Manufacturing

The explanations above have shown that Cellular Manufacturing as operational system for lean machining can be used as leverage for flexibility. The positive effects have also been proven in practice by Abele et al. (2013a). However, as stated before, flexibility considerations and actions always also need to be seen in an economic context. A commercial company's intended primary purpose is generating revenues. Thus, the costs per unit are still the most important financial measure for the profitability of the production system.

There are two major sources of influence on the operational costs of a Cellular Manufacturing system: the human and the technical capacity utilization (Bechtloff 2013). Consequently, the quality of the balancing of the technical system is of particular importance for the financial attractiveness. Its grade is depending on the requirements from clamping of the workpiece and from its orientation and location tolerances that have impact on the separability of the work content. The following formulas are proposed as KPIs for the quality of the balancing and for the utilizations to prepare a financial evaluation of Cellular Manufacturing concepts (Metternich et al. 2013a, b; Abele et al. 2013b).

$$\text{Average utilization of machine tools in the system} = \eta_m = \frac{\sum t_m}{n \times t_{\text{takt}}}$$

$$\text{Average utilization of operators in the system} = \eta_p = \frac{\sum t_p}{m \times t_{\text{takt}}}$$

$$\text{Machine cost ratio} = k = \frac{n \times c_m}{n \times c_m + m \times c_p}$$

$$\text{Cell efficiciency of the system} = \eta_c = k \times \eta_m + (1 - k) \times \eta_p$$

$$\text{Line balancing efficiency} = \mu = \frac{\sum t_m}{n \times (t_{\text{takt}} - t_{wc})}$$

with

c_m	Machine-hour rate without labour costs
c_p	Labour costs per hour
m	Number of operators
n	Number of machine tools
t_m	Standard time machining (complete machining)
t_p	Standard time manual processes incl. paths
t_{takt}	Takt time
t_{wpc}	Standard time work piece change

Versus the complete machining of workpieces on one machine, the financial calculation of Cellular Manufacturing concepts benefits from a parallelization of work being executed on several machine tools at the same time. The capacity utilization losses resulting from a suboptimal balancing and the increased human capacity needs in Cellular Manufacturing can thus be offset by a reduced overall operational time of the system. The main reasons are that operations are parallelized and multiple-machine operation by only one worker can easily be implemented. Based on these considerations, Bechtloff (2013) conducts a static cost comparison of complete machining and Cellular Manufacturing, taking human, technical and organizational parameters into account.

Bechtloff (2013) proves that Cellular Manufacturing is economically attractive under certain conditions, depending on the setups of both the complete machining reference scenario and the Cellular Manufacturing line design. The reference scenario is mainly determined by the number of machine tools that are operated by one operator. The Cellular Manufacturing line predominantly depends on the capital investment costs of the machinery being used. The break-even point can generally be expressed via the overall machining time of the workpiece. Considering a reference scenario of two machines being operated by one operator in complete machining, for example, a comparing Cellular Manufacturing line with up to four right-sized machine tools is economically preferred for workpieces with a machining time beyond 4 min.

Further research efforts in this area aim on an increased financial attractiveness through a reduction of the manual work via automation. Again, the capital investment is limited, leading to low-cost automation solutions (Seifermann et al. 2014).

Summary and Outlook

The value creating supply chain of physical products involves several actors of production, logistics and traffic and is thus highly divers and volatile. Disturbances in one discipline can easily affect the others. This chapter focuses on the possibilities flexibility in production offers for dealing with fluctuations from logistics and traffic. In order to get an impression of the flexibility requirements, the chapter starts with mapping the supply chain and describing the nature and severity of potential disturbances using the fault-tree analysis method. Currently, most production systems are decoupled from logistics and traffic by excessive inventory buffers. Thus, the effects of fluctuations on production from these disciplines are absorbed at the costs of overcapacity in people, inventory, processes and machinery. The popular concept of lean production identifies and addresses these overcapacities as "waste" and was therefore taken as a basis for further improvement discussions in the chapter.

After having developed the flexibility needs, a closer look at the term "flexibility" itself in manufacturing contexts as well as on the different types of flexibility was taken. Volume and mix flexibility were identified as the types most relevant for industrial companies. Thus, different models for measuring volume and mix flexibility in the literature have been reviewed. The chapter then turned to the actions for increasing flexibility in production systems and presented selected technical, organizational and human approaches. The focus was from then on set to lean production systems and its special conditions and prerequisites. First, selected approaches for addressing volume and mix flexibility in lean assembly systems were introduced. Namely, these are an adaptation of the output levels of an assembly to expected demand scenarios (takt times) via different predefined and tested operational setups and the SMED method for systematically reducing changeover time and thus enabling mix flexibility. The concept of Chaku-Chaku cells was introduced.

Whereas flexibility in lean assembly systems is quite widespread, lean machining systems still lag behind. Possible explanations for this fact were discussed, before the concept of Cellular Manufacturing as an approach to transfer the takting and balancing from assembly to machining was introduced. In Cellular Manufacturing lines, a complex machining task, usually executed on one machine tool in complete machining, is split into several fragments being distributed to a number of smaller, specialized and cheaper machine tools adequate for the operation (right-sized equipment). Volume and mix flexibility advantages due to the Cellular Manufacturing concept were presented. As any flexible concept is only viable in industrial practice if it is also economically advantageous, the promising results of Bechtloff's (2013) in-depth static cost comparison of Cellular Manufacturing versus complete machining were introduced.

Future research in this area should consider the following:

- Quantify the actual flexibility potential of Cellular Manufacturing with an adequate evaluation model.
- Further evaluate the link of inventories that currently decouple production, logistics and traffic to flexibility, e.g. clarify the question of how much inventory can be reduced by the introduction of Cellular Manufacturing, etc.
- Further push the economic boundaries of Cellular Manufacturing by identifying low-cost automation and quality possibilities.
- Link the flexible Cellular Manufacturing concept closer with internal logistics as key interface to the external logistics and traffic.

References

Abele, E., Bechtloff, S., Cachay, J., & Eichhorn, N. (2010). Schlanke Zerspanung. Qualifizierung zur Optimierung von Zerspanprozessen. *Werkstatt + Betrieb wb, 9*, 94–98.

Abele, E., Bechtloff, S., & Krause, F. (2011). Flexible Serienfertigung im Kundentakt. *Werkstatt + Betrieb, 6*, 24–27.

Abele, E., Bechtloff, S., & Seifermann, S. (2012). Sequenzfertigung für flexible und schlanke Zerspanung. *Productivity Management, 17*(1), 45–48.

Abele, E., Bechtloff, S., Seifermann, S., & Törnig, W. (2013a). Flexibilitätssteigerung durch Lean Production für die Zerspanung. *VDI-Z, 155*(4), 53–55.

Abele, E., Metternich, J., Bechtloff, S., & Heller, J. (2013b). Sequenzfertigung im Performance-Check. *Werkstatt + Betrieb, 4*, 32–36.

Abele, E., Kuhn, S., & Hueske, B. (2008a). Overall equipment flexibility. *Zeitschrift für wirtschaftlichen Fabrikbetrieb ZWF, 103*(5), 322–327.

Abele, E., Rumpel, G., & Kuhn, S. M. (2008b). Flexible Produktionskonzepte für die saisonale Produktion. *Zeitschrift für wirtschaftlichen Fabrikbetrieb ZWF, 103*(9), 585–588.

Aberle, G. (2003). *Transportwirtschaft. Einzelwirtschaftliche und gesamtwirtschaftliche Grundlagen.* München: Oldenbourg.

Askar, G. (2008). *Optimierte Flexibilitätsnutzung in Automobilwerken.* Dissertation. Shaker, Aachen.

Bateman, N., Stockton, D., & Lawrence, P. (1999). Measuring the mix response flexibility of manufacturing systems. *International Journal of Production Research, 37*(4), 871–880.

Bellmann, K., Himpel, F., & Böhm, A. (2009). Messung von Flexibilität in der Produktion. In J. Strohhecker & A. Größler (Eds.), *Strategisches und operatives Produktionsmanagement. Empirie und Simulation* (pp. 221–240). Wiesbaden: Gabler.

Beach, R., Muhlemann, A. P., Price, D. H. R., Paterson, A., & Sharp, J. A. (2000). A review of manufacturing flexibility. *Theory and Methodology. European Journal of Operational Research, 122*, 41–57.
Beamon, B. (1999). Measuring supply chain performance. *International Journal of Operations and Production Management, 19*(3), 275–292.
Bechtloff, S. (2013). *Identifikation wirtschaftlicher Einsatzgebiete der Sequenzfertigung in der Bohr- und Fräsbearbeitung von Kleinserien.* Aachen: Shaker.
Beck, S. (2012). *Entwicklung und Validierung einer Systematik zur Flexibilitätsanalyse und -bewertung von Produktionslinien bei einem Befestigungshersteller.* Thesis. TU Darmstadt.
Bertsche, B., & Lechner, G. (2004). *Zuverlässigkeit im Fahrzeug- und Maschinenbau: Ermittlung von Bauteil- und Systemzuverlässigkeiten.* Stuttgart: Springer.
Boysen, N., Fliedner, M., & Scholl, A. (2007). A classification of assembly line balancing problems. *European Journal of Operational Research, 183*, 674–693.
Browne, J., Dubois, D., Rathmill, K., Sethi, S., & Stecke, K. (1984). Classification of flexible manufacturing systems. *Flexible Manufacturing Systems Magazine, 2*, 114–117.
Burbidge, J. L. (1975). *The introduction of group technology.* New York: John Wiley & Sons.
Chryssolouris, G. (2006). *Manufacturing systems: Theory and practice.* New York: Springer.
Chryssolouris, G., & Lee, M. (1992). An assessment of flexibility in manufacturing systems. *Manufacturing Review, 5*, 105–116.
Cummings, T., & Blumberg, M. (1987). Advanced manufacturing technology and work design. In T. D. Wall, W. Clegg, & N. J. Kemp (Eds.), *The human side of advanced manufacturing technology* (pp. 37–60). Chichester, NY: Wiley.
Das, S. (1996). The measurement of flexibility in manufacturing systems. *The International Journal of Flexible Manufacturing Systems, 8*, 67–93.
De Toni, A., & Tonchia, S. (1998). Manufacturing flexibility: A literature review. *International Journal of Production Research, 36*(6), 1587–1617.
Drew, J., McCallum, B., & Roggenhofer, S. (2005). *Unternehmen Lean. Schritte zu einer neuen Organisation.* Frankfurt: Campus.
Ehmer, H. (2009). Arbeitszeitgestaltung zwischen Arbeit auf Abruf und Vertrauensarbeitszeit. *Industrie Management, 25*(3), 10–12.
ElMaraghy, H. A. (2005). Flexible and reconfigurable manufacturing systems paradigms. *International Journal of Flexible Manufacturing Systems (IJFMS), 17*(4), 261–276.
Erlach, K. (2010). *Wertstromdesign. Der Weg zur schlanken Fabrik.* 2., bearbeitete und erweiterte Auflage. Berlin: Springer.
Faißt, J. (1992). Instrumente des Arbeitszeitmanagements. In C. Scheeweiß (Ed.), *Kapazitätsorientiertes Arbeitszeitmanagement* (pp. 41–60). Heidelberg: Physica.
Gerwin, D. (1989). Manufacturing flexibility in the CAM era. *Business Horizons, 32*(1), 78–84.
Gudehus, T. (2010). *Logistik. Grundlagen – Strategien – Anwendungen.* 4., aktualisierte Auflage. Heidelberg: Springer.
Gupta, Y. P., & Goyal, S. (1989). Flexibility of manufacturing systems: Concepts and measurements. *European Journal of Operational Research, 43*, 119–135.
Hackstein, R. (1989). *Produktionsplanung und -steuerung (PPS). Ein Handbuch für die Betriebspraxis.* Zweite, überarbeitete Auflage. Düsseldorf: VDI-Verlag.
Halubek, P. (2012). *Simulationsbasierte Planungsunterstützung für Variantenfließfertigung.* Dissertation. Vulkan, Essen.
Hüttmeir, A., De Treville, S., Van Ackere, A., Monnier, L., & Prenninger, J. (2009). Trading off between heijunka and just-in-sequence. *International Journal of Production Economics, 118*, 501–507.
Imai, M. (2001). *KAIZEN. Der Schlüssel zum Erfolg im Wettbewerb.* Munich: Econ.
Irani, S. A. (1999). *Handbook of cellular manufacturing systems.* New York: John Wiley & Sons.
Jack, E. P., & Raturi, A. (2002). Sources of volume flexibility and their impact on performance. *Journal of Operations Management, 20*, 519–548.

Jalizi, B., Korff, D., & Rost, R. (2009). Alleskönner oder Teamplayer: Mehrtechnologiemaschinen versus Mehrmaschinenkonzepte. *Werkstatt + Betrieb, 10*, 12–17.
Jünemann, R., & Schmidt, T. (2000). *Materialflußsysteme. Systemtechnische Grundlagen*. Berlin: Springer.
Kaluza, B., & Blecker, T. (2005). Flexibilität. State of the Art und Entwicklungstrends. In B. Kaluza & T. Blecker (Eds.), *Erfolgsfaktor Flexibilität* (pp. 1–28). Berlin: Erich Schmidt.
Koste, L. L., Malhotra, M. K., & Sharma, S. (2004). Measuring dimensions of manufacturing flexibility. *Journal of Operations Management, 22*, 171–196.
Krafcyk, J. (1988). Triumph of the lean production system. *Sloan Management Review, 30*(1), 41–52.
Krüger, A. (2004). *Planung und Kapazitätsabstimmung stückzahlflexibler Montagesysteme*. München: Utz.
Lane, G. (2007). *Made-to-order. Excelling in a high-mix, low-volume environment*. New York: Productivity Press.
Lanza, G., Rühl, J., & Peters, S. (2009). Bewertung von Stückzahl- und Variantenflexibilität in der Produktion. *Zeitschrift für wirtschaftlichen Fabrikbetrieb ZWF, 104*(11), 1039–1044.
Leavitt, H. J. (1965). Applied organizational change in industry: Structural, technological and humanistic approaches. In J. G. March (Ed.), *Handbook of organizations* (pp. 1144–1170). Chicago, IL: Rand McNally.
Liker, J. K. (2013). *Der Toyota-Weg. 14 Management-Prinzipien des weltweit erfolgreichsten Automobilkonzerns*. Munich: Finanzbuchverlag.
Lotter, B., Hartel, M., & Menges, R. (1995). *Manuelle Montage – wirtschaftlich gestalten. Neuzeitliche Analysen und Planungsmethoden zur Montagerationalisierung*. Renningen-Malmsheim: expert
Martin, H. (2014). *Transport- und Lagerlogistik. Planung, Struktur, Steuerung und Kosten von Systemen der Intralogistik*. 9., vollständig überarbeitete und aktualisierte Auflage. Wiesbaden: Springer.
Matzka, J., Di Mascolo, M., & Furmans, K. (2009). Buffer sizing of a Heijunka Kanban system. *Journal of Intelligent Manufacturing*. doi:10.1007/s10845-009-0317-3 (Published online Sept 2009).
Metternich, J., Bechtloff, S., & Seifermann, S. (2013). Efficiency and economic evaluation of cellular manufacturing to enable lean machining. In *46th CIRP Conference on Manufacturing Systems 2013. Procedia CIRP* (Vol. 7, pp. 592–597).
Metternich, J., Böllhoff, J., Seifermann, S., & Beck, S. (2013). Volume and mix flexibility evaluation of lean production systems. In *Procedia CIRP 9 (2013), 2nd CIRP Global Web Conference* (pp. 79–84).
Milling, P., & Thun, J.-H. (2005). Steigerung der Flexibilität in Produktionsprozessen durch intergrative Fertigung. In B. Kaluza & T. Blecker (Eds.), *Erfolgsfaktor Flexibilität – Strategien und Konzepte für wandlungsfähige Unternehmen* (pp. 191–208). Berlin: Schmidt.
Mitrofanov, S. P. (1966). *The scientific principles of group technology* (Vol. I, II & III). Yorkshire: National Lending Library for Science and Technology.
Nyhuis, P., Reinhart, G., & Abele, E. (2008). *Wandlungsfähige Produktionssysteme*. Berlin: PZH Produktionstechnisches Zentrum.
Ohno, T. (1993). *Das Toyota-Produktionssystem*. Frankfurt: Campus.
Parker, R., & Wirth, A. (1999). Manufacturing flexibility: Measures and relationships. *European Journal of Operations Research, 118*, 435–438.
Pfeiffer, W., & Weiß, E. (1994). *Lean Management. Grundlagen der Führung und Organisation lernender Unternehmen*. 2., überarbeitete und erweiterte Auflage. Berlin: Erich Schmidt.
Pfohl, H.-C. (2010). *Logistiksysteme. Betriebswirtschaftliche Grundlagen*. 8., neu bearbeitete und aktualisierte Auflage. Heidelberg: Springer.
Richter, M. (2006). Gestaltung der Montageorganisation. In B. Lotter & H.-P. Wiendahl (Eds.), *Montage in der industriellen Produktion. Ein Handbuch für die Praxis* (pp. 95–126). Berlin: Springer.

Schlick, C., Bruder, R., & Luczak, H. (2010). *Arbeitswissenschaft*. 3., vollständig überarbeitete und erweiterte Auflage. Heidelberg: Springer.
Schmeißer, T. R. (2012). *Maßnahmen zur Erweiterung der Mengenflexibilität und ihre Auswirkungen in der industriellen Produktion*. Thesis. TU Darmstadt.
Schulze, T. (2012). *Erstellen eines Modells der Produktion, inklusive zugehöriger Supply Chain und Untersuchung der flexiblen Reaktionsmöglichkeiten auf mögliche Störgrößen*. Thesis. TU Darmstadt.
Seifermann, S., Böllhoff, J., Metternich, J., & Bellaghnach, A. (2014). Evaluation of work measurement concepts for a cellular manufacturing reference line to enable low cost automation for lean machining. In *Proceedings of the 47th CIRP Conference on Manufacturing Systems. Procedia CIRP* (Vol. 17, pp. 588–593).
Sekine, K. (1995). *Produzieren ohne Verschwendung. Der japanische Weg zur Schlanken Produktion*. Landsberg: Moderne Industrie.
Sende, C., Galais, N., Moser, K., & Hasenau, K. (2011). Flexibler Mitarbeitereinsatz. Ergebnisse einer deutschlandweiten Unternehmensbefragung. *Industrie Management, 27*(4), 52–56.
Sethi, A., & Sethi, S. (1990). Flexibility in manufacturing: A survey. *The International Journal of Flexible Manufacturing Systems, 2*, 289–328.
Shazly, S. (2011). *Development and process optimization of existing SAP-tools for inventory control and inventory reduction with particular consideration of traffic-related aspects*. Thesis. TU Darmstadt.
Shingo, S. (1989). *A study of the Toyota production system from an industrial engineering viewpoint* (Revised Edition). Cambridge, Norwalk: Productivity Press.
Spengler, T., Volling, T., & Rehkopf, S. (2005). Chaku Chaku Systeme in der Montage konsumtennaher Erzeugnisse. Eine Fallstudie bei Rahmenauftragsfertigung. In H.-O. Günther, D. C. Mattfeld, & L. Suhl (Eds.), *Supply Chain Management und Logistik. Optimierung, Simulation, Decision Support* (pp. 249–276). Heidelberg: Physica.
Stölzle, W., Heuser, K. F., & Karrer, M. (2004). *Erfolgsfaktor Bestandsmanagement. Konzept – Anwendung – Perspektiven*. Zürich: Versus.
Suarez, F., & Cusumano, M. (1996). An empirical study of manufacturing flexibility in printed circuit board assembly. *Operations Research, 44*, 223–240.
Takeda, H. (1996). *Das System der mixed production. Flexibel – rationell – kundenorientiert*. Landsberg: Moderne Industrie.
Takeda, H. (2009). *Das synchrone Produktionssystem. Just-in-time für das ganze Unternehmen*. 6., aktualisierte Auflage. Munich: Finanzbuch Verlag.
Thums, A. (2004). *Formale Fehlerbaumanalyse*. Dissertation. Augsburg.
Trist, E. L., & Bamforth, K. W. (1951). Some social and psychological consequences of the longwall method of coal-getting. *Human Relations, 4*, 3–38.
Ulich, E. (2011). *Arbeitspsychologie*. 7., neu überarb. und erw. Aufl. Zürich: vdf-Hochschulverlag.
Van Hop, N. (2004). Approach to measure the mix response flexibility of manufacturing systems. *International Journal of Production Research, 42*(7), 1407–1418.
Wahab, M. (2005). Measuring machine and product mix flexibilities of a manufacturing system. *International Journal of Production Research, 43*(18), 3773–3786.
Zelenovic, D. (1982). Flexibility—A condition for effective production systems. *International Journal of Production Research, 20*(3), 319–337.
Zelinski, P. (2006). Why Boeing is big on right-size machine tools. *Modern Machine Shop, 78*(10), 72–79.

Simulation-Based Assessment of Lean Production Methods: Approaches to Increase Volume and Variant Flexibility

Markus P. Roessler, Felix Wiegel, Eberhard Abele, and Joachim Metternich

Introduction

In today's globalized economic world, product cycles are becoming shorter and trends are spreading quicker throughout industries than ever before. Shifting and more challenging customer demands pose a challenge for manufacturing companies' goals for efficiency and target-optimized production systems. The production coordination costs especially increase due to broader variant spectrums and smaller batch sizes (Wiendahl et al. 2004). To avoid such a complexity trap in the long term, companies must introduce measures to increase their flexibility, not only in terms of applied equipment, but most significantly through flexible logistics and workforce (Günther 2007).

In Germany, these developments show an increased impact due to the relatively high proportion of the manufacturing industry at the country's gross value added. This export driven system, with an export rate of 40 % in 2013 which amounts to 1.094 trillion euros of exports (Statistisches Bundesamt 2014), is scattered over more than 6000 manufacturing companies. Ninety nine per cent of these make up the often-cited backbone of the German industry, consisting mostly of small and medium-sized companies which have positioned themselves with highly specialized niche products as worldwide market leaders in their respective business sector (Abele and Reinhart 2011). In order to keep their competitive advantages and

M.P. Roessler (✉) • F. Wiegel • E. Abele • J. Metternich
Institute of Production Management, Technology and Machine Tools, Technische Universität Darmstadt, Otto-Berndt-Strasse 2, 64287 Darmstadt, Germany
e-mail: roessler@ptw.tu-darmstadt.de; wiegel@ptw.tu-darmstadt.de; abele@ptw.tu-darmstadt.de; metternich@ptw.tu-darmstadt.de

E. Abele et al. (eds.), *Dynamic and Seamless Integration of Production, Logistics and Traffic*, DOI 10.1007/978-3-319-41097-5_5

defend their world leading role, it is key for these companies to adapt to highly flexible production systems as well as to meet efficiency requirements (Bornewasser and Zülch 2012).

With the implementation of lean production systems, companies have found a portfolio of tools to tackle these modern challenges, mainly by reducing waste or non-value-adding process steps such as waiting periods, unnecessary processes, or high levels of work in process (WIP) due to overproduction (Dickmann 2009). In order to study and train the effects of these methods, the Technische Universitaet Darmstadt has founded the Centre for Industrial Productivity (CiP), which provides a realistic production environment (Abele et al. 2007). In this context, an extensive simulation model of the CiP has been developed by the team of researchers and enhanced and analyzed mainly by Busse (2013), allowing inexpensive validation of new methods and problem-solving approaches.

This chapter's objective is to exploit the simulation environment of CiP under the scope of flexibility. Thus, chosen methods of lean production are implemented into the model and analyzed according to their effects on two flexibility indicators as well as the overall production performance. Three main factors of the production system will be addressed: equipment, employees, and internal logistics. The complete approach aims to identify measures to increase volume and variant flexibility especially for producing companies and thus to react quicker to external market requirements. It enables a future consideration of other supply chain participants like logistics contractors and transportation system providers and the impact of the applied methods to these external entities.

The following chapter covers the literature background for flexibility, the application to the production environment, and its current developments. After that the applied research methodology, the description of the simulation model, applied lean production methods (production leveling, flexible manufacturing systems, and demand managed milk run), and the testing scenarios will be described. Every method will be applied individually and compared to the basic system configuration. Finally, a conclusion is drawn, arguing the different methods' capability to achieve a more flexible production system.

Approaches for the Description of Flexibility

Generically speaking, flexibility describes the ability to adapt to external influences, though different academic approaches have tailored their understanding of flexibility to their purposes (Möslein-Tröppner 2010). In this context, a various number of definitions are available; see Kaluza and Blecker (2005) and Van Hop (2004). Due to manufacturing systems, flexibility is the capability of the components and processes to adapt to changes of input factors, processes, or output demand (Bornewasser and Zülch 2012). A different approach describes flexibility as the extent in which the characteristics of a system allow a reversible adaptation (Westkämper and Roscher 2005). Therefore, a system can be considered more flexible if change processes can be adopted with minimal decreases in efficiency, or if the range of system conditions in which efficient change processes are possible is broader. This implies that flexibility cannot only be seen as the ease of change but also involves the costs by acknowledging efficiency as a key component. In other words, flexibility is the capability of a system to adjust quickly and with low additional costs to changing constraints within a specified corridor (Schellmann 2012).

The economic reality increasingly considers flexibility as a significant factor for maintaining competitive advantages, especially for high tech industries such as in Germany. A recent, nationwide study of 1221 companies links flexibility with product quality and shows its ability to serve as a unique selling point (Jeske et al. 2011). The major challenge is to implement system flexibility without reducing cost efficiency, e.g., due to the necessity of additional resources or employees (Corsten and Gössinger 2009). It can be argued though that the advantages of meeting customer demand more closely and thus building up a competitive advantage compensate for the costs of implementation on a monetary level. Hence, a well-balanced production strategy cannot be obtained without including flexibility towards external factors (Bornewasser and Zülch 2012).

A closer look at flexibility in the manufacturing context shows a wide range of types of flexibility which can be identified. Flexibility indicators vary from singular input factors like machines or employees to long-term and company-wide factors. While figures focus on different aspects of the manufacturing system, an overall structure can be derived, showing the dependencies and levels of flexibility, and allowing to identify factors which influence flexibility.

Extensive overviews about the different kinds of flexibility are provided by the articles of Browne et al. (1984) as well as Sethi and Sethi (1990). With respect to the production focus of this chapter, volume and variant flexibility have been chosen as the key factors (Zhang et al. 2003). The basic characteristics of flexibility—rapid, reversible, and cost-effective adaptations—are still valid. In addition, the indicators chosen can be directly linked to customer satisfaction and thus help achieve the stated advantages of flexibility.

Volume flexibility represents the capability to efficiently and economically change the production volumes (Browne et al. 1984; Yamashina et al. 1986). The necessity of changing the number of produced units results from the changing of surrounding conditions (e.g., change in demand) and should be reversibly adaptable, even without operation interruption (Jack and Raturi 2002).

Variant flexibility describes the capability to efficiently produce different product types or variants. This definition is similar to the one in Browne et al. (1984). The term *variant flexibility* is used according to Yamashina et al. (1986). This flexibility type implies a production system's stability towards the dropping and adding of products or product variants from the production sequence, while leaving the remaining products unaffected (Rogalski 2009). Therefore, variant flexibility is mainly based on a detailed design of the production processes as well as the equipment used (Schellmann 2012).

In order to understand the origins of these flexibility indicators and to put their effects into context, a structural framework must be deployed. Koste and Malhorta (1999) developed a five-tier structure that stems from the organization structure of a company. In this logical structure, higher tiers are influenced by the lower tiers.

Volume and variant flexibility in this respect are tier-3 flexibilities on production site level. The basis of the structure is built up by machine, material flow, and employee flexibility. They make up the individual input factors of the production process and thus it can be easily understood how they have significant impacts on volume and variant flexibility. Since these are the only dependencies of interest within the scope of this chapter, the complexity of this model can be severely reduced.

In accordance with the work of Zhang et al. (2003), one can distinguish levels of flexibility within the production process. Flexible manufacturing capability is measured by volume and variant flexibility. This generates the ability to produce according to ever-changing customer demands. The potential flexibilities of machines, employees, material flow, and routes affect this ability directly and exclusively, thereby defining the flexible manufacturing competence as a basis. Hence, any ambitions to change the volume or variant flexibility depend on pulling the levers of machine, employee, material flow, and route flexibilities (Zhang et al. 2003).

In order to alter certain flexibility types on the level of flexible manufacturing competence, one can identify a set of adjustment enablers. These adjustment enablers describe crucial system properties such as (Hernandez Morales 2003; Roscher 2008):

- *Ability to enhance or reduce system features*, which states the system's ability to compensate layout changes of capacity factors, e.g., machines, employees, or other resources.
- *Neutrality of function or use*, which describes the ability of resources to serve multiple functions.
- *Modularity*, which defines standardized interfaces to allow a modular change of objects.
- *Ability to interlink as well as to integrate and to disintegrate*, which defines the capability of cost-efficient rearrangements of system connections.
- *Mobility*, which allows the free movement of production factors such as employees moving to the workstation where needed.

The derived enablers give an understanding of the interdependencies of flexibility within a manufacturing process. They act as input variables, which can be used to influence volume and variant flexibility through the discussed set of factors.

Methodology and Research Approach

Production Environment

In order to evaluate a systematic deployment of lean production methods towards their impact on the introduced production flexibility measures, the research presented in this chapter utilizes a simulation-aided approach. The model refers to the CiP. This educational facility for lean production methods represents a real production environment and its value stream in detail. Two main departments are divided into machining and assembly and can be set up according to various production scenarios in order to study and teach them (Brungs 2012). Both departments are linked by a logistics concept allowing, amongst others, the introduction of pull coordination by Kanban. This environment represents an ideal setting to validate improvement methods because of the broad portfolio of processes available and the realistic combination of machining and assembly areas (Abele et al. 2011a).

The output product of the facility is a *pneumatic cylinder*, which is being produced and assembled from raw material and a set of semifinished parts. The first step of the value stream is machining and this is divided into five stages. Aluminum profiles are cut into slices by a saw and set up in the CNC milling machine to produce two types of cylinder bottoms (BP). Simultaneously, the CNC lathe produces four different types of piston rods (PR) from sawed raw material in one step. The variant types differ by diameter and thread type (inner and outer) and can include any combination of these characteristics. This differentiation directly corresponds to the variant options of the final product: possible diameters are Ø25 or Ø40 mm (D24, D40) with inner (I) or outer (O) threads as a connection for the piston rod to the respective setup. All the parts from machining pass through the final two stages, washing and quality control, on a FIFO basis before they are handed over to the assembly supermarket.

The assembly area is set up as a U-cell. Plastic pistons, lids, housings, gaskets, and screws are supplied by a shipping warehouse as outsourced parts and are put together with the in-house manufactured parts to produce the final cylinder. Before shipping, all finished cylinders have to pass a final functional test. The material flow can be organized by different logistics concepts. The spectrum varies from self-supply by assembly workers to a route-based time-dependent milk run concept, exempting workers from non-value-adding activities (Abele et al. 2011b). The simulation model has been created out of the value stream map of the CiP (see Fig. 1).

Simulation Modeling and System Constraints

To enable a quantification of effects to external entities, the simulation of a real production environment must not be limited to an enclosed production plant and direct suppliers and customers must be included. Thus, a scenario for the CiP is derived based on the work of Kunert (2013) and Busse (2013).

The scenario places the CiP production plant as a tier-1 manufacturer of an automotive supply chain. Therefore, tier-2 manufacturers act as direct suppliers for the production of pneumatic cylinders, and the customer takes the role of an OEM. Customers are modeled as black boxes, only providing necessary input parameters

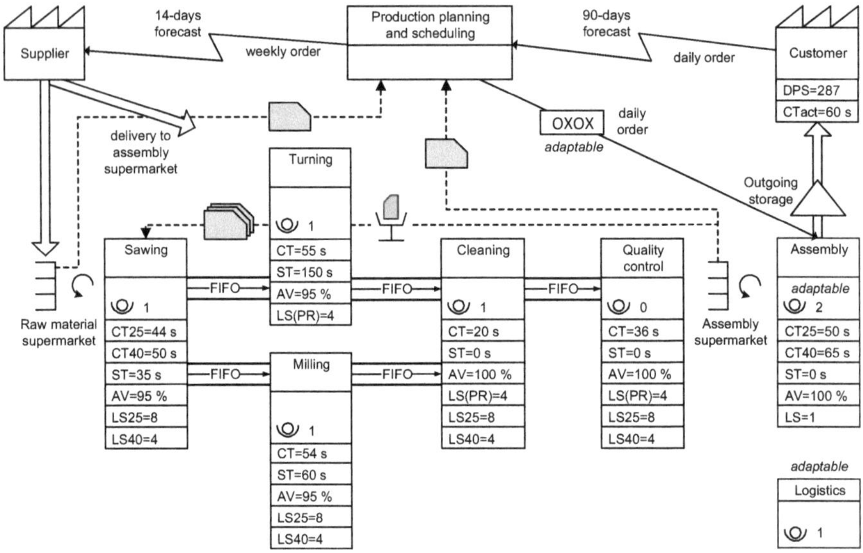

Fig. 1 Value stream of CiP's production system, where leveling, logistics management, and assembly workers; count are adaptable, cf. (Roessler et al. 2014)

for the tier-1 manufacturer CiP. In addition, suppliers for raw material are considered in the simulation coping with any given demand. The complete setting allows the evaluation of the impact of lean production methods to the specific production system as well as in- and outbound logistics.

The product mix at CiP covers four variants in all considered scenarios:

- Pneumatic cylinder Ø25 mm with inner thread (D25-I)
- Pneumatic cylinder Ø25 mm with outer thread (D25-O)
- Pneumatic cylinder Ø40 mm with inner thread (D40-I)
- Pneumatic cylinder Ø40 mm with outer thread (D40-O)

The simulation model was realized using the software AnyLogic 6. Additionally, a 3D-CAD model of the CiP helps to simulate employee movement. This feature allows the assessment of logistics routes and assures the plausibility of the sequence of processes (see Fig. 2). There the main output factors of the simulation are shown, the produced parts and their respective variants, the overall worker count in the scenario, the simulation time in seconds, the covered distance of the logistician in meters (CD), the load of the logistician in percent, and the amount of inventory in different areas of the system. Not shown in this figure is the mask for parameterizing scenarios and the capacity utilization of each entity.

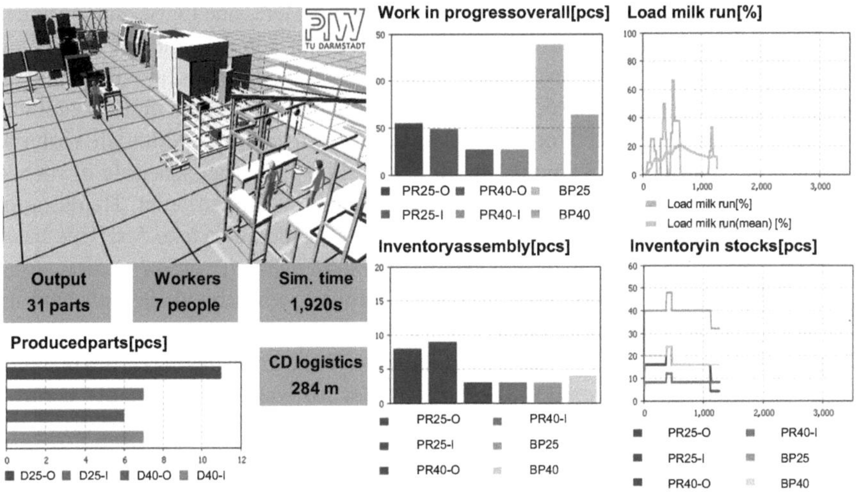

Fig. 2 Graphical user interface of the material flow simulation model with 3D visualization and chosen output performance indicators

Material Flow Model and Crucial Parameters

Customer Demand and Production Management

Following the idea of a pull-oriented production, the description of the material flow usually starts with the customer demand. Its average time allowance of 60 s per order of one cylinder sets the upper limit for average cycle times within the CiP. The simulation of CiP's information flow is also widely pull oriented; i.e., the demand of the downstream workstation regulates upstream production or assembly by Kanban. Customer demand and production control are realized through the pacemaker process, represented by the assembly (U-line) of the cylinders. Kanban cards, based on assembly demand, control the machining stages for sawing, turning, and milling, as well as supplier delivery of outsourced products. Raw materials are provided on a daily basis and in sufficient quantity.

In addition to the customer demand, supply of outsourced parts by external suppliers provides another input parameter for the simulation. Since supply is set to cover demand at any time given as a simplification, it will not be discussed further at this point. The number of products manufactured by the CiP represents the central output parameter.

Logistics Concept

In accordance with the lean production approach of separating value-adding from non-value-adding work (Womack et al. 1990), the material transport between all workstations at CiP is realized by a milk run concept. Here, one logistician provides material for the workstations. In contrast to a direct supply concept, orders are bundled instead of being executed individually. Fixed routes are designed in accordance with production demand at certain stations at which empty containers are collected and replaced on the basis of Kanban control. Although material supplied is controlled by demand, the milk run follows a strict schedule of routes. Thus, every route starts at a predetermined time and in a fixed order. Handling of the small load carriers (SLC) at every station is considered to take 5 s. This basic concept implemented is called *milk run with time-dependent control*.

The model for simulation differentiates two routes, whereby the logistic activity is not limited to only supply the assembly but also the machining department. At first, on route 1 parts are collected from the individual workstations of the machining department and delivered to the FIFO buffer for washing and quality control. Production at the machining department is indirectly initiated by the arrival of the milk run, which is therefore acting as the trigger for parts production. Subsequently, the milk run merges parts from machining and external suppliers into the central assembly supermarket as part of route 1. Supply to the assembly line is covered by route 2. Kanban cards control the demand and trigger the delivery amounts from the supermarket provided for each workstation, where empty containers are collected and returned (Busse 2013; Kunert 2013).

The entire production plant uses a set of standardized SLC for transportation. The carrier's small dimensions of 243 mm × 162 mm × 129.5 mm combined with an average weight of about 1.00 kg enable manual handling. Each carrier with one Kanban card holds eight parts for the Ø25 mm variant at any production or assembly stage, and four in the case of the Ø40 mm variant (parameter LS). Exceptions are made for piston rods, which are transported in a set of four, regardless of their characteristics. Supermarket capacities for assembly are limited to four carriers per part, while the assembly line itself can hold up to two carriers per part. Finally, the milk run cart can store up to 24 SLC.

Assembly and Machining

The complete cylinder assembly (see Fig. 1) can be broken down into a five-stage process (five workstations), which by default is covered by *two workers*. Worker 1 is responsible for the first and last station, while worker 2 takes care of the remaining three workstations. Assembly starts by joining piston rod and piston at workplace (WP) 1. Then the bottom and the lid are added and screwed to the casing at WP2 and WP3. Quality control then checks the final products for functionality at WP4 before packaging prepares them for shipment at WP5. Employee assignment can be varied for assembly, which will be subject to a closer analysis later. As the pacemaker process, assembly sets the production rate for all upstream workstations. Its cycle time therefore matches the customer's tact closely at an average of 55 s per cylinder.

For machining, every workstation, i.e., sawing, turning, milling, cleaning/quality control, is covered by one worker. The Kanban cards from assembly arrive at a batch size planning board to form effective batch sizes for machining with low and standardized changeover times already in the basic scenario. Thus, manufacturing indirectly is triggered by Kanban through the milk run. Turning, milling, cleaning, and quality control work on a FIFO basis, their processes are triggered by material arrival from the respective upstream station. Raw material is stocked in a raw material supermarket before sawing.

Process Parameters of the Simulation

This paragraph shows the main influencing parameters for simulation; this covers product mix, cycle times, machine setup, batch sizes, employee count, and customer demand.

The simulation of customer demand is realized by generating a randomly modeled demand. The basic setup for the *random unleveled overall demand* is based on a Bernoulli distribution, with an expected product mix ratio of 2:1 for diameter (Ø25–Ø40 mm) and 1:1 for thread type (inner to outer). Per shift, the sum of individual demands cumulates to 287 cylinders in random order (DPS = demand per shift), which corresponds to a customer demand rate of 60 s per part (CTact).

Setup times, which occur at every production changeover to another product variant, are also incorporated in the simulation. Their implementation is realized by raw material distinction linked to a production delay. Changeover and production (firstly at the saw) starts if a batch size of four Kanban cards for a specific variant is reached. Setup times for the machining area are:

- 150 s turning machine setup time
- 60 s milling machine setup time
- 35.25 s saw setup time

The average cycle times (CT) for all processes (machining and assembly) are:

CT turning machine	55 s/part
CT milling machine	54 s/part
CT saw	44 s/part (Ø25 mm variant)
	50 s/part (Ø40 mm variant)
CT cleaning	20 s/part
CT quality control	36 s/part
CT assembly	50 s/part (Ø25 mm variant)
	65 s/part (Ø40 mm variant)
	55 s/part (basic mix of 2:1)

Each workstation at CiP's machining area has its own worker, except for cleaning and washing, which is covered by one worker for both stations. In addition, assembly is staffed with two workers for the standard scenario and one logistician for the whole plant:

- Four workers for machining
- Two workers for assembly
- One worker for logistics

Input factors, like raw material or outsourced parts, are assumed to cover any demand given at any time. The simulation allows for adaptation of all listed parameters through input masks. Therefore, a variety of scenarios can be implemented to analyze the system response.

Identification of Flexibility Enhancing Lean Production Methods

Lean production offers a wide range of approaches and methods promising an impact on volume and variant flexibility. Therefore, only a selection of the most promising ones will be analyzed for the scope of this chapter. At the same time, the chosen methodology of using the mentioned simulation environment is a limiting factor for the applicable methods.

Since the simulation closely represents the CiP setup, positioning and number of machines, storage areas, and workstations as well as the overall production layout and space are taken as given and cannot be changed. Thus, approaches such as using highly flexible machining centers or oversized production capacities will not be analyzed at this point because this usually is associated with high investments. Though, the capacity fixation only applies to machining. Adaptation of labor resources, however, is possible in a relatively short time with low investment and is expected to provide significant leverage on volume flexibility as well as, to a certain extent, on variant flexibility, due to the labor-intensive assembly process. Hence a *flexible manufacturing system* will be implemented for analyses in the simulation (see Fig. 3).

Preliminary studies show that a time-dependent milk run relies on a stable variant mix. It is important to note that this is a matter of management rather than an inherent characteristic of the chosen supply system (Kilic et al. 2012). Therefore, *demand managed control of the milk run* will be analyzed as a measure towards higher variant flexibility.

The final method of *production leveling* has been chosen due to its central role within the concept of lean production. According to Syska (2006) and Wuthnow and Deuse (2009), it is expected to provide a general improvement of production performance including variant and volume flexibility. As a result, the structure in Fig. 3 demonstrates the selected methods and their expected influence on volume and variant flexibility.

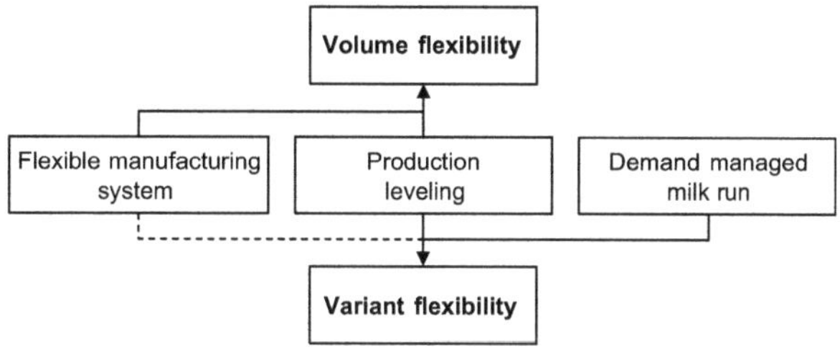

Fig. 3 Structure of expected lean production methods' qualitative impact on flexibility indicators, cf. (Busse 2013)

Production leveling

Production leveling embodies one of the key aspects of lean production and, thus, was already implemented to the real production plant CiP. However, it is expected that leverage on volume as well as variant flexibility can be examined by utilizing the simulation model.

The basic idea of production leveling is analyzing the random input of customer demand over a defined period of time and then converting it into a repetitive pattern for every product or product variant (Buhl et al. 2009). Adjustments to production patterns are made by varying the parameters production volume, mix, and sequence (Dickmann 2009; Erlach 2010). The main advantage, therefore, is enabling a solid planning and thus leveling out material demand and capacity use throughout the entire value stream (Liker 2004). Its main characteristics are:

- Standardized production process and steadily predictable workload
- Balanced utilization of production and logistics capacities
- Constant material and information flow
- Minimization of stocks throughout the entire value stream
- Establishment of stability over the entire production system
- Ability to react quicker to customer demand due to shorter lead times

Flexible Manufacturing System

For the scope of this chapter, the term *flexible manufacturing system* clusters a set of human resource measures. Generally, manufacturing flexibility with respect to employees can be achieved by adapting a new shift system or varying the total number of employees (Sende et al. 2010). The objective of these measures is to cope with changing customer demand while maintaining high resource productivity. Therefore, they have an obvious impact on volume flexibility and in certain scenarios—depending on production line design—on variant flexibility as well. In order to successfully implement this system, highly trained employees, as well as a flexible working time model, are needed (Sende et al. 2011).

Measures like this will only be executed efficiently if the production lines are suitable, work content is highly standardized, and employees with broad qualifications are available. Major advantages of this method are listed below (Takeda 2005):

- Fast reaction to fluctuations of demand
- Steady and high work productivity
- Minimized operator application and minimal part transport

Demand Managed Milk Run

The final method analyzed is a variation of milk run management, the so-called *demand managed milk run*. So far milk run activity is scheduled by predetermined times and routes. The time aspect is replaced by demand triggered Kanban control, while routes are maintained. Therefore, drifting material demand due to variations of product mix can be flexibly coped with (Brungs 2012).

Time-based control is based on the assumption of a steady and determined product mix. Changes hold the risk of under- or overloading the milk run and generally causing an unbalanced workload. Demand managed milk run control uses minimum storage limits as trigger to redirect milk run activity. Thus, time intervals between deliveries vary in accordance with the material use. The supply quantity, however, is fixed.

Scenarios and Simulation Setup

A set of four sub-scenarios is implemented to analyze system performance, which differ in their product mix. Thus, allowing conclusions towards variant flexibility for the implemented production setup. Volume flexibility can be analyzed by adapting output amounts in every scenario accordingly.

Each test run of the simulation is based on $n = 20$ production cycles in order to compensate for any initial distortions. While output amounts can be adopted gradually to test for volume flexibility, product mix and thus system performance towards variant flexibility are tested by distinction of the four sub-scenarios. Distinctions between scenarios are made by varying the relative amount of Ø40 mm cylinders from 0 to 100 % in four steps. Since thread types have marginal impact on the simulation—they only differ by setup times—no changes will be made to their distribution, which is fixed to 50:50 %. The sub-scenarios are:

- Sub-scenario 1:
  - Distribution diameter Ø25:Ø40 mm = 100:0 % = 1:0
- Sub-scenario 2:
  - Distribution diameter Ø25:Ø40 mm = 66:33 % = 2:1
- Sub-scenario 3:

  - Distribution diameter Ø25:Ø40 mm = 33:66 % = 1:2
- Sub-scenario 4:
  - Distribution diameter Ø25:Ø40 mm = 0:100 % = 0:1

Sub-scenario 2 is the setup the processes in CiP originally were designed for. Also the average customer's tact (see Fig. 1) is calculated based upon this sub-scenario. To enable a measurement of the methods' impacts on different performance indicators, one basic scenario of the CiP has to be defined. This basic scenario is described by the previous paragraphs and can be summed by the following:

- Production control: Four sub-scenarios, thereby randomly distributed variants; customer orders directly control the assembly line, no leveling
- Employee setup: 4 workers for machining, 2 for assembly, one logistician
- Logistics concept: Milk run with time-dependent control

Application to the Simulation Model

Production Leveling

The performance impact of the method will be measured by direct comparison to the system's basic setup in the four sub-scenarios. *Production leveling* is realized by an implemented program code, which results in the following production patterns for the variants, triggering assembly (see Fig. 1):

- For sub-scenario 1:
  A–B–A–B–A–B–A–B–A–B–A–B–A ...
- For sub-scenario 2:
  A–B–A–B–C–D–A–B–A–B–C–D–A ...
- For sub-scenario 3:
  C–D–C–D–A–B–C–D–C–D–A–B–C ...
- For sub-scenario 4:
  C–D–C–D–C–D–C–D–C–D–C–D–C ...

where A = Ø25 mm, inner thread; B = Ø25 mm, outer thread, C = Ø40 mm, inner thread; D = Ø40 mm, outer thread.

Simulation results for the output rate hereby show a marginal increase of 1 % in the exemplary case of scenario 2 (see Table 1). At the same time, other performance indicators, such as cycle time, WIP, and non-value-adding work ratio, are reduced by a similar degree. Standard deviation for output rate was diminished by 94 % compared to the normal setup.

Figure 4 maps the output rate to the cumulated output of one test run for sub-scenario 2. It is shown that leveled production slightly outperforms its

Table 1 Performance for basic and leveled configurations of sub-scenario 2

Performance indicator	Basic configuration	Leveled configuration
Simulation time (s)	17,283	17,068
Output rate (mean) (pcs/min)	0.96	0.97
Output rate (max) (pcs/min)	1.00	1.01
Standard deviation output rate (pcs/min)	0.0070	0.0004
Value-adding time (%)	72.73	73.61
Non-value-adding time (transport) (%)	7.52	7.51
Waste (waiting, motion) (%)	19.76	18.88
WIP overall (pcs)	272	267

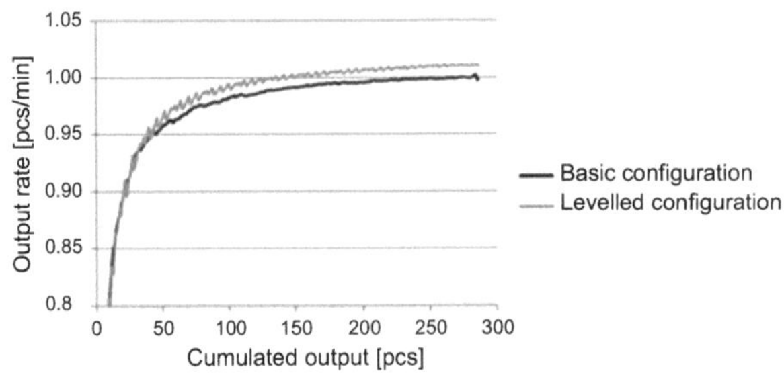

Fig. 4 Output rates for basic and leveled configurations of scenario 2

non-leveled counterpart for any output, though the effect tends to diminish for small outputs. The leveled production's output rate shows a characteristic zigzag-shaped trend. Its origin can be found in the fixed production pattern stated above under consideration of variant-dependent cycle times.

Results for the remaining sub-scenarios follow this general pattern. No sub-scenario shows a decline in a performance indicator after introduction of leveling. This can be explained through the existence and use of the batch size planning board for the machining area, which arranges standardized and stable changeover times. Results show only minor differences between the basic and leveled configuration, which also could occur from random fluctuations in the simulation model. So the impact on volume flexibility nearly can be neglected for the case of CiP (see Table 2).

Mapping this interrelation by approximating a continuous course over the variants gives an idea of leveling's leverage towards variant flexibility (see Fig. 5). Hereby the variant mix was slowly increased from 0 to 100 % proportion of the D40 variant.

Improvements brought about by this method also to the variant flexibility are quite marginal, but the system generally works more stable. The more homogenous production patterns smooth out material supply fluctuations within the CiP by reducing peak demands caused by randomly incoming customer's orders. In addition, lower standard deviation for performance indices indicates a more predictable and reliable planning. The impact of production leveling for variant and volume flexibility is close to zero in the case of CiP.

Table 2 Comparison of output rates by sub-scenario for $n = 20$ runs

Experiment configuration	Basic configuration	Leveled configuration
Output rate (mean) (pcs/min) for sub-scenario 1	0.95	0.95
Output rate (mean) (pcs/min) for sub-scenario 2	0.96	0.97
Output rate (mean) (pcs/min) for sub-scenario 3	0.77	0.80
Output rate (mean) (pcs/min) for sub-scenario 4	0.52	0.53

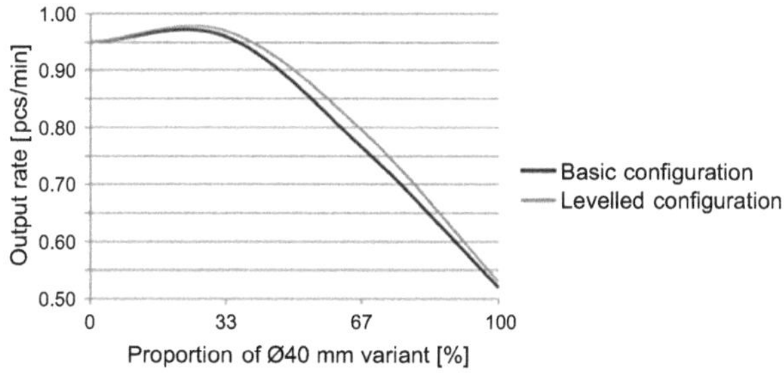

Fig. 5 Impact of variant mix on output rate for scenario 2

Flexible Manufacturing System

The *flexible manufacturing system's* main objective is to enable volume flexibility, i.e., guarantee the ability to meet customer demand. Hence, the number of assembly workers will be adapted, while fixing the product mix and the logistics concept to the basic scenario (and using sub-scenario 2 for product mix). Since the assembly line consists of five workstations, scenarios for a maximum of five employees are discussed. For each setting of employees, the whole work content for assembly has to be redistributed and balanced according to the worker count. Results are represented in Table 3.

Considering the average output rate, adding employees to the manufacturing cell gradually increases the system performance (output rate). This effect however is limited to an amount of five workers, because no additional workplaces are available. More than five workers result in high shares of waste (waiting time). When increasing the number of workers beginning with one, WIP levels decrease because very few parts have to wait for the next process and directly are handed over to the next worker. The indicators for value-adding activities and waste both reach their optimum for two assembly employees, unsurprisingly the state for which this assembly line was designed for. Greater amounts of workers cannot fully exploit the added resources.

It is important to acknowledge that conclusions so far are based on the average output rate. However, mapping the trend of output rate to cumulated output reveals significant findings (see Fig. 6). It is shown that after overcoming initial distortions, output rates tend to level out for scenarios up to three assembly employees. Scenarios with 4 or 5 employees, on the other hand, reach a maximum of close to 1.6 parts per minute before output rates decrease linearly. Apparently, the upstream

Table 3 Performance for different assembly configurations of sub-scenario 2

Performance indicator	One worker	Two workers (basic)	Three workers	Four workers	Five workers
Simulation time (s)	33,593	17,283	15,608	14,858	14,425
Output rate (mean) (pcs/min)	0.50	0.96	1.12	1.42	1.44
Output rate (max) (pcs/min)	0.52	1.00	1.16	1.65	1.66
Standard deviation output rate (pcs/min)	0.003	0.007	0.030	0.086	0.0561
Load milk run (%)	44.28	52.62	53.54	55.32	55.67
Load assembly employees (%)	98.73	96.14	70.19	55.41	45.25
Value-adding time (%)	50.64	72.73	65.80	59.23	54.04
Non-value-adding time (transport) (%)	7.38	7.52	6.69	6.15	5.57
Waste (waiting, motion) (%)	41.98	19.76	27.50	34.62	40.40
WIP overall (pcs)	421	272	226	207	200

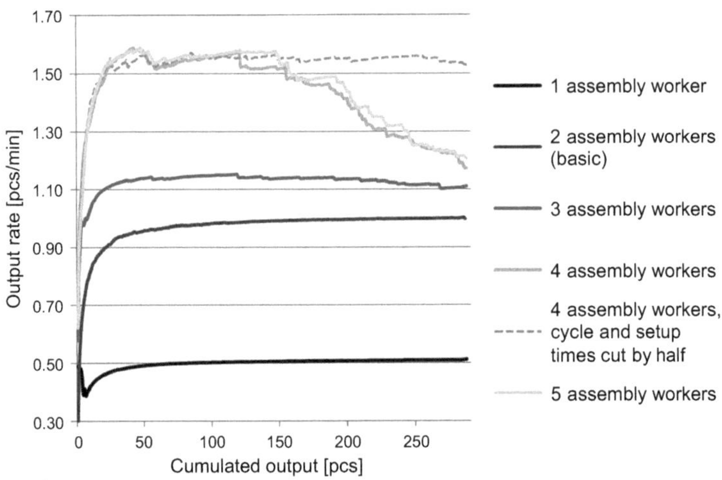

Fig. 6 Impact of assembly workers' count on output rate for sub-scenario 2

processes in the machining department are unable to support the viable output rate of the assembly line in the long run. Compensating for this effect, i.e., enhancing machining productivity by decreasing cycle and setup times, leads to the expected, stable output rates, as shown by the dotted line in Fig. 6.

Testing for variant flexibility is realized by scrutinizing the introduced variant mix sub-scenarios for different employee intensities. Since the assembly line setup seems to favor 2 and 4 employees in terms of performance indices, these will be tested exclusively. Table 4 generally confirms the findings from above, where higher employee counts achieve better output rates. On average, the output rate could be increased by 22 % over all sub-scenarios. The gap between the options largely depends on the customer product mix, with greater impact for sub-scenarios

Table 4 Comparison of output rates by employee settings

Experiment configuration	Two assembly workers (basic)	Four assembly workers
Output rate (mean) (pcs/min) for sub-scenario 1	0.95	1.26
Output rate (mean) (pcs/min) for sub-scenario 2	0.96	1.42
Output rate (mean) (pcs/min) for sub-scenario 3	0.77	0.85
Output rate (mean) (pcs/min) for sub-scenario 4	0.52	0.54

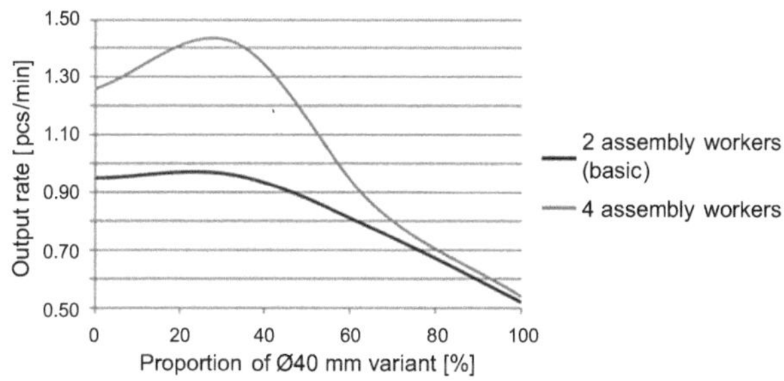

Fig. 7 Impact of variant mix on output rate for sub-scenario 2

favoring the Ø25 mm variant (sub-scenario 1 and 2). Again, mapping this interrelation by approximating a continuous course shows the potency of flexible manufacturing systems depending on product mix (see Fig. 7). For higher ratios (>75 %) of the Ø40 mm variant, almost no performance improvement can be realized through flexible manufacturing system application.

Demand Managed Milk Run

The final method examined is the *demand managed milk run*. Instead of following a fixed route schedule, milk run activity is triggered by demand. The performance impact is measured by applying the method to the four sub-scenarios. The results are shown in Table 5.

The values need to be set into context to allow a conclusion, i.e., compared with the time-based milk run management (see Table 6). Confirming the expectations, demand-based milk run management has an overall positive effect on system performance throughout all sub-scenarios considered. The effect is related to the product mix, this time showing greater impact for higher ratios of the Ø40 mm variant.

Table 5 Performance results for demand managed milk run and all sub-scenarios

Performance indicator	Sub-scenario 1	Sub-scenario 2	Sub-scenario 3	Sub-scenario 4
Simulation time (s)	17,453	17,201	21,695	25,342
Output rate (mean) (pcs/min)	0.95	0.96	0.82	0.71
Output rate (max) (pcs/min)	0.99	1.00	0.89	0.80
Standard deviation output rate (pcs/min)	0.005	0.009	0.016	0.044
Load milk run (%)	98.50	97.60	98.40	98.20
Value-adding time (%)	69.01	75.77	67.78	61.55
Non-value-adding time (transport) (%)	14.07	13.95	14.05	14.04
Waste (waiting, motion) (%)	16.92	10.28	18.17	24.41
WIP overall (pcs)	339	294	371	412

Table 6 Comparison of output rates by milk run management method

Experiment configuration	Milk run with time-dependent control (basic)	Demand managed milk run
Output rate (mean) (pcs/min) for sub-scenario 1	0.95	0.95
Output rate (mean) (pcs/min) for sub-scenario 2	0.96	0.96
Output rate (mean) (pcs/min) for sub-scenario 3	0.77	0.85
Output rate (mean) (pcs/min) for sub-scenario 4	0.52	0.71

It can be stated that the method applied enables the milk run to show higher variant flexibility. This is confirmed by mapping the interrelation of output rate to product mix through approximating a continuous course (see Fig. 8).

A closer look at the milk run workload reveals that a demand-driven approach obviously results in higher initial capacity utilization. However, transport amounts under demand-based management smoothen out in the path of the test run and tend to reach lower capacity utilization through homogenization of transported quantities. Besides this important advantage, the maximum transport capacity is not exceeded at any time or for any sub-scenario.

The major downside of this approach is visualized by the figures for WIP. For the basic scenario (and sub-scenario 2) with a demand managed milk run, there are 294 parts in process compared to 272 for the respective time-based milk run. The gap of 22 parts is explained by the more frequent transport cycles, which tend to drag more material into the process.

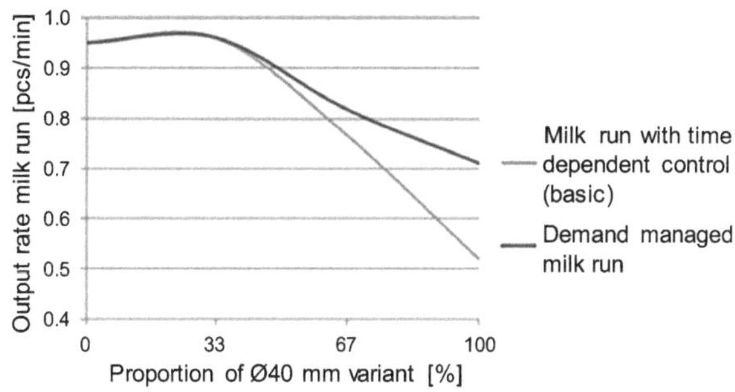

Fig. 8 Impact of variant mix on output rate of milk run for all sub-scenarios (mean values, for each sub-scenario $n = 20$ runs)

Conclusions and Outlook

Today's manufacturing industry faces challenges from its increasingly dynamic and globalized market environment. Continuous efficiency improvements or waste reductions closely linked to the introduction of lean production methods are their direct answer. Hence, reacting quickly to customer's demands while maintaining high efficiency becomes an increasingly important competitive factor. The implementation and performance impact of this feature of flexibility has been the scope of this paper. It has been shown that production flexibility can be analyzed by volume and variant flexibility using material flow simulation.

The research in this chapter focused on three lean production methods, namely, production leveling, flexible manufacturing system, and demand managed milk run. While all methods showed performance improvement potential, i.e., in terms of stability, performance, and variant and volume flexibility, a final assessment must include the financial expenditures of implementation. With regard to the methods applied, flexible manufacturing systems involve severe prerequisites and therefore costs, while leveling a production or changing the milk run management generally adds up to a minor amount. This chapter indicates the leverage of all tested methods towards the selected flexibility indices.

As the results show, flexible manufacturing systems have proven to be the most potent tool to improve volume flexibility. The variant flexibility of the CiP production system on the other hand was proven to be successfully enhanced through demand managed milk run control. Compared to the results of the flexible manufacturing system and demand managed milk run control, production leveling showed less compelling results in the applied production setting. However, the standard deviation of the output rate could be significantly reduced. So a more stable system output could be achieved. Due to its core role in lean production, leveling can be seen as an enabler for further value stream improvements, and its minor costs should also be considered as a positive factor.

Another key finding is that merging the analyzed methods holds significant synergy potential. In a further test setting, all three methods were applied simultaneously to the simulation model (four assembly workers, leveling, and demand management of milk run). Over all four sub-scenarios the average output rate could have been enhanced by 25 % to better fulfill customers' demand.

In terms of product mix, each method shows a characteristic effective range, where its potential for performance improvement is the highest (see Figs. 5, 7, and 8). Therefore, enhancing production flexibility is only achievable by exploiting a combination of tools. While these statements are also true for volume flexibility, they are limited to a certain output range. Results show that one-sided approaches will push other segments of the production process past their limits. Hence, the path for improving flexibility must be pursued on a value stream level with a set of methods.

This also paves the way for further analysis of flexibility measures, e.g., relating to flexible machining or new concepts of automation. Supplementation of the given simulation model represents the next step towards an integrative flexible production environment, covering flexible machines, flexible employees, flexible logistics, and a flexible planning. This chapter also shows that material flow simulation is a suitable tool for measuring the effects of a lean production implementation. Using different scenarios, single methods as well as a combination of methods can be examined with regard to their effectiveness with relatively low investments. Further research potential besides implementing toolsets is seen in analyzing the impact of flexibility enhancing measures to external supply chain participants like logistics contractors and transportation system providers.

References

Abele, E., Anderl, R., Brungs, F., & Mosch, C. (2011a). Materialflusssimulation in der schlanken Produktion [Material flow simulation in lean production]. *Productivity Management, 16*(4), 48–51.

Abele, E., Bechtloff, S., & Krause, F. (2011b). Flexible Serienfertigung im Kundentakt [Flexible serial production by customers tact]. *Werkstatt und Betrieb, 6*, 24–27.

Abele, E., Eichhorn, N., & Brungs, F. (2007). Mitarbeiterqualifikation in einer realen Produktionsumgebung [Employee qualification in a real production environment]. *Zeitschrift für wirtschaftlichen Fabrikbetrieb, 102*(1–2), 741–745.

Abele, E., & Reinhart, G. (2011). *Zukunft der Produktion: Herausforderungen, Forschungsfelder, Chancen [Future of production: Challenges, fields of research, opportunities]*. München: Carl Hanser Verlag.

Bornewasser, M., & Zülch, G. (2012). *Arbeitszeit—Zeitarbeit: Flexibilisierung der Arbeit als Antwort auf die Globalisierung [Working time—temporary work: Flexible work as a response to globalization]*. Wiesbaden: Springer Gabler.

Browne, J., Dubois, D., Rathmill, K., Sethi, S. P., & Stecke, K. E. (1984). Classification of flexible manufacturing systems. *FMS Magazine*, 747–756.

Brungs, F. (2012). *Der Milkrun in der Produktionslogistik [Milk run in production logistics]*. Dissertation. Technische Universität Darmstadt, Darmstadt.

Buhl, M., Bohnen, F., Deuse, J., & Schneider, R. (2009). Effiziente Kleinserienfertigung durch Produktionsnivellierung [Efficient production of small batches through balancing]. *Productivity Management, 14*(4), 19–22.

Busse, A. (2013). *Erweiterung eines Simulationsmodells zur Bewertung intralogistikbezogener Maßnahmen hinsichtlich Produktivität und Flexibilität [Enhancement of a simulation model for evaluation of intralogistics-related measures regarding prodctivity and flexibility]*. Thesis. Academic advisor: Roessler, M. P. Technische Universität Darmstadt, Darmstadt.

Corsten, H., & Gössinger, R. (2009). *Produktionswirtschaft: Einführung in das industrielle Produktionsmanagement [Production economics: Introduction to industrial production management]*. München: Oldenbourg Verlag.

Dickmann, P. (2009). *Schlanker Materialfluss [Lean material flow]*. Berlin: Springer.

Erlach, K. (2010). *Werstromdesign: Der Weg zur schlanken Fabrik: Value stream design: The way to the lean factory*. Berlin: Springer.

Günther, W. A. (2007). *Neue Wege in der Automobillogistik: Die Vision der Supra-Adaptivität [New ways in automotive logistics]*. Berlin: Springer.

Hernandez Morales, R. (2003). *Systematik der Wandlungsfähigkeit in der Fabrikplanung [Nomenclature of mutability in factory planning]*. Dissertation. Leibnitz Universität Hannover, Düsseldorf: Fortschritt-Berichte VDI.

Jack, E., & Raturi, A. (2002). Sources of volume flexibility and their impact on performance. *Journal of Operations Management, 20*(5), 519–548.

Jeske, T., von Garrel, J., & Starke, J. (2011). Erfolgsfaktor Flexibilität: Ergebnisse einer deutschlandweiten Unternehmensbefragung [Success factor flexibility: Results of a German industrial survey]. *Industrial Engineering, 64*, 20–23.

Kaluza, B., & Blecker, T. (2005). Flexibilität: State of the Art und Entwicklungstrends [Flexibility: State of the art and trends]. In B. Kaluza & T. Blecker (Eds.), *Erfolgsfaktor Flexibilität: Strategien und Konzepte für wandlungsfähige Unternehmen*. Berlin: Erich Schmidt Verlag.

Kilic, H. S., Durmusoglu, M. B., & Baskak, M. (2012). Classification and modeling for in-plant milk run distribution systems. *International Journal of Advanced Manufacturing Technology, 62*(9–12), 1135–1146.

Koste, L., & Malhorta, M. (1999). A theoretical framework for analyzing the dimensions of manufacturing flexibility. *Journal of Operations Management, 18*(1), 75–93.

Kunert, C. (2013). *Entwicklung eines Simulationsmodells zur Bewertung intralogistikbezogener Maßnahmen hinsichtlich Produktivität und Flexibilität [Development of a simulation model for evaluation of intralogistic-related measures regarding productivity and flexibility]*. Thesis. Academic advisor: Roessler, M. P. Technische Universität Darmstadt, Darmstadt.

Liker, J. K. (2004). *The Toyota way: 14 management principles from the world's greatest manufacturer*. New York: McGraw-Hill.

Möslein-Tröppner, B. (2010). *Produktionswirtschaftliche Flexibilität in Supply Chains mit hohen Absatzrisiken [Commercial production flexibility in supply chains with high sales risks]*. Bamberg: University of Bamberg Press.

Roessler, M. P., Metternich, J., & Abele, E. (2014). Learning to see clear: Quantification and multidimensional assessment of value stream mapping alternatives considering variability. *Business and Management Research, 3*(2), 93–109. doi:10.5430/bmr.v3n2p93.

Rogalski, S. (2009). *Entwicklung einer Methodik zur Flexibilitätsbewertung von Produktionssystemen: Messung von Mengen-, Mix- und Erweiterungsflexibilität zur Bewältigung von Planungsunsicherheiten in der Produktion*. Dissertation. Universität Karlsruhe, Karlsruhe.

Roscher, J. (2008). *Bewertung von Flexibilitätsstrategien für die Endmontage in der Automobilindustrie [Evaluation of flexibility stratgies for assembly shops in the automotive industry]*. Dissertation. Universität Stuttgart, Stuttgart.

Schellmann, J. H. (2012). *Bewertung kundenspezifischer Mengenflexibilität im Wertschöpfungsnetz [Evaluation of customer specific volume flexibility in value added network]*. München: Herbert Utz Verlag Wissenschaft.

Sende, C., Galais, N., & Moser, K. (2011). Flexibler Mitarbeitereinsatz: Ergebnisse einer deutschlandweiten Unternehmensbefragung [Flexible worker deployment: Results of a German industrial survey]. *Industrie Management, 27*, 52–56.

Sende, V., von Garrel, J., & Tackenberg, S. (2010). Strategien des flexiblen Personaleinsatzes [Strategies for flexible worker deployment]. *Personalwirtschaft, 10*(2010), 27–29.
Sethi, A. K., & Sethi, S. P. (1990). Flexibility in manufacturing: A survey. *International Journal of Flexible Manufacturing Systems, 2*, 289–328.
Statistisches Bundesamt. (2014). Exportquote [Export rate]. *Statistisches Bundesamt*. Accessed August 1, 2014, from https://www.destatis.de
Syska, A. (2006). *Produktionsmanagement [Production management]*. Wiesbaden: Gabler Verlag.
Takeda, H. (2005). *Das synchrone Produktionssystem: Just-in-Time für das ganze Unternehmen [The synchrone production system: Just-in-time for the whole enterprise]*. München: Redline.
Van Hop, N. (2004). Approach to measure the mix response flexibility of manufacturing systems. *International Journal of Production Research, 42*(7), 1407–1418.
Westkämper, E., & Roscher, J. (2005). Bewertung flexibler Endmontagesysteme für die Automobilindustrie durch Simulation des Realbetriebs [Evaluation of felxible assembly systems in the automotive industry using simulation]. *Werkstattstechnik online, 95*(4), 181–185.
Wiendahl, H., Gerst, D., & Keunecke, L. (2004). *Variantenbeherrschung in der Montage [Coping with variants in assembly]*. Berlin: Springer.
Womack, J. P., Jones, D. T., & Roos, D. (1990). *The machine that changed the world: The story of lean production*. New York: Free Press.
Wuthnow, A., & Deuse, J. (2009). Systematische Nivellierung von Wertströmen [Systematic leveling of value streams]. *Zeitschrift für wirtschaftlichen Fabrikbetrieb, 104*(4), 229–234.
Yamashina, H., Okamura, K., & Matsumoto, K. (1986). Flexible manufacturing systems in Japan: An overview. In *Proceedings of the 5th International Conference on Flexible Manufacturing Systems* (pp. 405–416), London, November 1986.
Zhang, Q., Vonderembse, M., & Lim, J. (2003). Manufacturing flexibility: Defining and analyzing relationships among competence, capability, and customer satisfaction. *Journal of Operations Management, 21*(2), 173–191.

Revenue Sharing Between Production, Logistics and Traffic: An Experimental Analysis of the Actors in Distribution Logistics

Ralf Elbert, Özhan Özsucu, and Cora Bogusch

Introduction

Due to the increasingly uncertain environmental developments, distribution logistics should explore ways to respond to the end customer demand with flexible systems (Zhang et al. 2005). An improvement of the distribution logistics system is feasible by considering the whole system including the interdependencies of production and transportation as relevant adjacent systems. For example, it is particularly necessary to link the operations of the logistics service provider (LSP) and the manufacturing company in order to provide an integrated supply chain (Mortensen and Lemoine 2008). Otherwise, the efficiency of the operations can be decreased. In addition to this, the required information across the supply chain actors should be integrated to achieve an improved holistic performance (Daugherty et al. 2002). However, the challenge we face today is to integrate the individual elements of the distribution logistics system including the interdependencies with production and traffic.

As the link between production and consumption of goods, logistics represents a crucial part of a holistic view regarding production, logistics and traffic (PLT) (Pfohl 2010). These integrative characteristics of logistics in general and logistics services in particular have to be analysed by studying the actors' behaviour. Logistics services are getting an increased importance in the global competition (Langley 2012; worldwide revenue contract logistics: USD542 billion (+6.8 %, 2009)). Outsourcing and the megatrend of globalization have led to a reduction in the depth of added value and extremely complex structures of logistics systems

R. Elbert (✉) • Ö. Özsucu • C. Bogusch
Management and Logistics, Technische Universität Darmstadt, Hochschulstraße 1, 64289 Darmstadt, Germany
e-mail: elbert@log.tu-darmstadt.de; oezsucu@log.tu-darmstadt.de; bogusch@log.tu-darmstadt.de

E. Abele et al. (eds.), *Dynamic and Seamless Integration of Production, Logistics and Traffic*, DOI 10.1007/978-3-319-41097-5_6

(Wallenburg and Wieland 2010). Therefore, many customers of logistics services tend to reduce the complexity of their logistics operations with the help of an LSP (Bradley 1995). A rising number of outsourcing of logistics services and an expansion of the range of logistics service offerings are a result of this development. So, a shift from component solutions to an integrated solution from a single provider is observed (Mortensen and Lemoine 2008). It is even more important for the LSPs to identify customer requirements at an early stage and to develop appropriate logistics services or to align logistics processes on customer—e.g. manufacturing or transport connection—requirements. Consequently, the consideration of contractual arrangements and incentive systems is becoming a major aspect in terms of logistics outsourcing (Selviaridis and Spring 2007; Marasco 2008). With regard to outsourcing developments described, there are often very complex tasks for the involved actors because of large contract volumes and long contract periods (Razzaque and Sheng 1998). Therefore, a careful preparation of the cooperation like an early participation of production into logistics processes is particularly important. Especially, the role of LSP, who serves as an intermediary on the interface of production and traffic, is becoming increasingly important and strategic role within the supply chain (Selviaridis and Spring 2007; Kumar et al. 2006).

In this chapter, we will focus on the business relationship (like e.g. outsourcing relationships) between PLT. The behavioural insecurities and information asymmetries are relationship risks which lead to uncertainties within a business relationship (Tsai et al. 2012). In this context, the deployment of information concerning internal processes is a very important factor for logistics to design their processes more efficiently and effectively (Mohr and Spekman 1994; Stank et al. 1996). For example, the trust of production in logistics has a positive effect on the operational performance of logistics (Knemeyer and Murphy 2004). The level of information disclosure and the incentive to reduce opportunistic behaviour is dependent on the collaborative value for the actors (Nahapiet and Ghoshal 1998).

To achieve the goal of an improved overall result by disclosing information, the different actors of PLT should cooperate with each other. Against this background, in this chapter, the behaviour of actors within the distribution logistics system will be examined by means of laboratory experiments. Distribution logistics is a crucial part of the supply chain. With regard to inventory management, they have to ensure the availability of products (production interface), and simultaneously as part of transportation handling, they have to distribute these products to the end customer (traffic interface) (Klug 2010). Especially, the willingness to provide information of the actors will be explored for a guaranteed revenue share. To develop suitable logistics solutions, it is still necessary to consider all aspects in the calculation of prices for the offered services. For this purpose, it is also important to take production and transportation processes into account. With appropriate concepts regarding pricing or revenue sharing, advantages for all actors can be achieved (Bellantuono et al. 2009).

The need to harmonize the interest of production, logistics and traffic beyond the borders of the particular fields results from the interdependencies between them.

It is therefore necessary for the involved actors to exchange information with each other to reduce the opportunistic behaviour or asymmetric information in their relationship. It is common that the individual companies try to maximize their profit despite the possibility to generate a higher overall profit by cooperating along the supply chain. A constitutional setting where both participants could have incentives to gain a higher total profit in the relationship can be established by a revenue-sharing contract (Yao et al. 2008a). The idea of revenue-sharing contracts is to set an incentive with a guaranteed revenue share to generate higher overall revenue (Yao et al. 2008a). So, the revenue-sharing concept (Anderson and Thomson 2000; Langley 2009) could serve as an incentive model for the cooperation of production, logistics and traffic.

The research gap contains the application of revenue sharing in the relation of production and logistics. The literature in this field focuses almost always on the relationship between manufacturing companies and retailers (Rhee et al. 2010). In particular, the empirical research of the actors' behaviour regarding information disclosure in this business relationship is missing in the literature. Therefore, the research question is:

> Can the disclosure of information be increased by a revenue-sharing contract in a business relationship between production, logistics and traffic?

The objective of the chapter is to investigate if revenue sharing in production, logistics and traffic relationships can evoke a higher level of information disclosure. To answer the research question, we briefly give a theoretical background on the application of laboratory experiments and revenue sharing. Following this, we describe the experiment design with the setting and the experiment procedure. Then, we will depict the results of the experiment and the analysis of the generated data. Finally, concluding remarks complete the chapter.

Research Background

Revenue-Sharing Contracts in Business Relationships

The focus of the chapter is the setting of actors in the area of production, logistics and traffic which are already in a business relationship. Within this relationship, the logistician fulfils the task assigned by the customer (here production) (Wilding and Juriado 2004). So, a very important point in the business relationship is the trust of the partners in the binding commitment they arranged. Knemeyer and Murphy noted that a higher level of trust in the relationship exhibits a significantly higher perception of the operational performance (Knemeyer and Murphy 2004) Reciprocal investments are often part of an effective logistics service relationship (Lambert et al. 1999).

In this context, revenue-sharing contracts can be used to share information for an enhanced performance (Rhee et al. 2010). Revenue-sharing contracts were first introduced in the video rental industry in the 1990s. Due to a lower purchase price including a simple reward agreement (revenue share) with the film producers, the rental company Blockbuster was able to stock more films. As a result, they increased their turnover and both of them benefited from the revenue-sharing agreement (Cachon and Lariviere 2001). In comparison, the familiar sounding gain sharing used in employment relationships comprises a variety of incentives for a performance-based reward (Welbourne and Mejia Gomez 1995). Mostly, the literature concerning revenue-sharing contracts is focused on the coordination of supply chains (Cachon and Lariviere 2005; Giannoccaro and Pontrandolfo 2004; Yao et al. 2008b). The main topic of the papers is the design of the contracts. Especially, the wholesale price and the retailers' profit quota are continuously addressed as the content of these contracts (Dana and Spier 2001; Giannoccaro and Pontrandolfo 2009). An obstacle to deploy the full capability of the revenue-sharing contracts is the reluctant disclosure of information (Giannoccaro and Pontrandolfo 2009). So, the information sharing among the fields of PLT is crucial in the business relationship. In this research field, numerical operation models are commonly used as the main method regarding revenue sharing (Rhee et al. 2010; Lakhal 2006). To describe the strategic choice behaviour, the game theory, in terms of e.g. Stackelberg, is also applied (Chakravarty and Werner 2011; Yao et al. 2008a). Even more important is that there is marginal empirical evaluation of the application of the revenue-sharing contracts in the literature. Mostly, the buyer and supplier relationship is analysed in the papers (Rhee et al. 2010; Cachon and Lariviere 2005). Regarding logistics, just one publication was ascertained, which deals with the revenue share (Liu et al. 2013).

Laboratory Experiments and the Classic Investment Game

In order to answer the above-mentioned research question, we apply a laboratory experiment. Laboratory experiments are a highly valuable approach to gain insights into actors' behaviour (Knemeyer and Naylor 2011). For this purpose, the investment game by Berg et al. (1995) is applied. Laboratory experiments allow a clear temporal separation of cause and effect. More important, they enable the exploration of effects, which are impossible to work out in field experiments (Siemsen 2011). The control is the main advantage of laboratory experiments (Katok 2011; Highhouse 2009). This basically means that laboratory experiments allow identifying cause and effect due to their high internal validity. The researcher has the ability to control the institution and the environment (Deck and Smith 2013; Knemeyer and Naylor 2011). Laboratory experiments are already successfully employed to investigate behavioural aspects in logistics and supply chain management (Knemeyer and Naylor 2011; Tokar 2010; Bazerman 2002).

A broad body of literature has widely used the investment game to explore the comprehension of trust (Füllbrunn et al. 2011; Malhotra 2004; Pillutla et al. 2003).

The influence of incentives (Lewicki et al. 2006), formal contracts (Malhotra and Murnighan 2002) and reciprocity (Pillutla et al. 2003) on trust are highlighted in previous studies with the help of the investment game. Concerning the investment game, the reciprocal behaviour implicates a positive correlation between the investments of the sender and the paybacks of the responder. In the investigations of the investment game, Fahr and Irlenbusch (2000) reveal that this behaviour is the result of trust in the norm of reciprocity. Primarily, the investment game examines trust and reciprocity in a bilateral relationship (Berg et al. 1995). The game is played in a double-blind procedure. Initially, the sender can send the responder a share of his initial endowment. The received amount of the responder will be tripled. Now, the responder can share the amount with the sender and send a share back or keep all the money. The basic assumption of the investment game is the homo oeconomicus paradigm. Regarding the forecast, the responder has his own interests and keeps the received amount from the sender. In contrast, the sender anticipates these intentions and sends a zero amount to the responder. As a result, both of them send nothing to each other. This also reflects the equilibrium of the game. Following results are shown in Berg et al. (Berg et al. 1995). The sender had forwarded an amount on an average of USD5.16, whereas the responder had sent back an amount of USD4.66. The idea of reciprocity is usually known by the participants (Berg et al. 1995). The introduction of the revenue-sharing element in the relationship should evoke this reciprocity. So, the actors in the relationship should disclose information to generate higher overall revenue (Narayanan and Raman 2004). The remarks regarding revenue sharing the following hypothesis are determined.

H: With the revenue share the overall profit of the actors will increase.

The following experiment design will describe the testing process of the hypothesis.

The Experiment Design

The experiment is based on the first introduction of the investment game by Berg (Berg et al. 1995). According to the research question, we have adapted the investment game of Berg in the following points. The first assumption that the investments are considered as information relies on Coricelli (Coricelli et al. 2006). They analyse the effects of introducing asymmetric information and expectations in the investment game. They also describe that the value of the initial endowment is common information. Therefore, we adapt the assumption of Coricelli that the initial endowment represents the corporate internal data of the sender. Another point is highlighted by Rigdon (Rigdon 2009) and Fahr and Irlenbusch (Fahr and Irlenbusch 2000). They point out that the sender has more specific information of the investments in the relation as the responder. Consequently, we start the game with different initial endowments for the sender and responder. The adapted

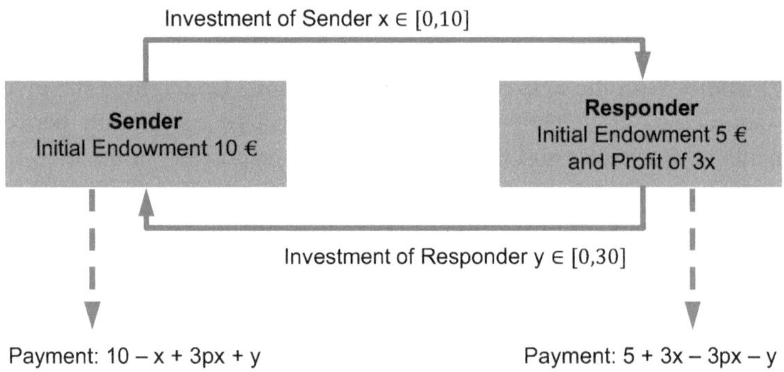

Fig. 1 The adapted investment game of Berg (Berg et al. 1995)

investment game lasts one round, a so-called one-shot game. Berg et al. reveal that a repeated game is not useful to discover trust (Berg et al. 1995, Fahr and Irlenbusch 2000; Houser et al. 2010). We focus on whether reciprocity is created by the treatment and not. Furthermore, Coricelli aims in his paper to test whether trust and reciprocity survive as patterns of behaviour in a setting where the personal decisions have low informational content to be cooperative. He achieves this by using asymmetric information setting in which only the trustee (responder) knows the size of the true surplus (Coricelli et al. 2006). Concerning this in our experiment design, the sender has no insights about the profit of the other actor due to information asymmetry, which was told to the participants. The sender has the certainty of a guaranteed revenue share (p) in the applied treatments. The guaranteed revenue share (p) of 20 % is based on the study conducted by Giannoccaro and Pontrandolfo 2009 (Giannoccaro and Pontrandolfo 2009). In addition, the responder has the opportunity to distribute everything over the guaranteed share to the sender. The sender receives an initial endowment of 10 MU (monetary units), whereas the responder receives an initial endowment of 5 MU. The alterations are shown in Fig. 1.

Experiment Procedure

Before the main experiment was conducted at the Technische Universität Darmstadt during the period of November, 2013, and January, 2014, we tested the experiment design with a pretest. Our pretest sample is composed of 12 participants, whereas in the main experiment, a total of 80 bachelor's or master's students participated in the experiment. The participants were students of different disciplines, but most of them were students of the departments law and economics. In this case, we play the experiment with 40 senders and 40 responders. The participants played the game once.

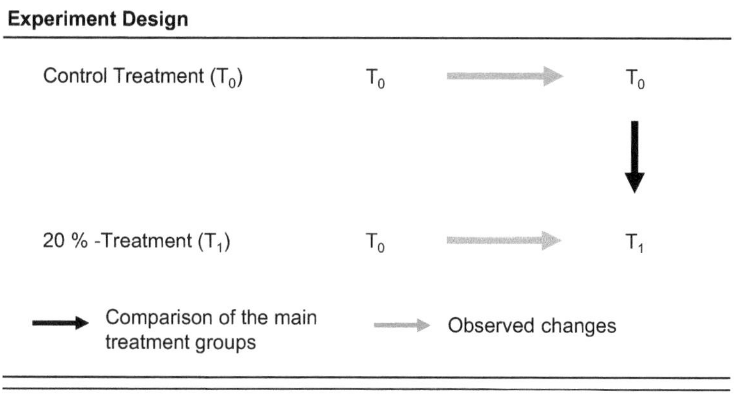

Fig. 2 Experiment design

The experiment design consists of two treatments. The control treatment T_0 and the 20 %-treatment T_1 (revenue-sharing treatment) in which the sender has the information that he gets a revenue share of 20 % of the multiplied amount he sends. A session is played in two separate periods and took less than 1 h to complete. The control treatment is applied to consider the randomization of the groups. Therefore, we play the control setting to ensure that all groups and their participants are homogeneous like similar experiences or identical educational backgrounds. After that we conduct the 20 %-treatment. The design of the experiment is depicted in Fig. 2.

At the beginning of the experiment, all participants are randomly assigned to the two roles—sender and responder. The experiment was conducted under anonymous conditions, i.e. participants did not know with whom they were matched. The experiment started after the experimenter read the instructions aloud to the group and answered questions. The sender can send any amount from 0 to 10 MU of his initial endowment to the responder. The responder receives the amount of the sender. The amount will multiply by a factor of three (Berg et al. 1995). The fact that the amount will multiply is known by the sender, but not the size of the true surplus. Only the responder knows the true value of the multiplier. Then the responder can send any amount of the received investment of the sender multiplied by a factor of three. After the first round is played, the pairs were reassigned with new partners, but they kept the same role. Then, the participants have played the corresponding treatment T_0 or T_1. Afterwards, the experiment ends. The experiment was conducted with the software z-Tree (Fischbacher 2007; Boero et al. 2009). In Fig. 3, the on-screen interface for the participants is presented from the viewpoint of the sender.

With the ending of the experiment, the participants complete a questionnaire. The objective of the questionnaire was to reveal whether the participants had any problems with the software or instructions. The evaluation of the questionnaire shows that the participants had no problems with both of them.

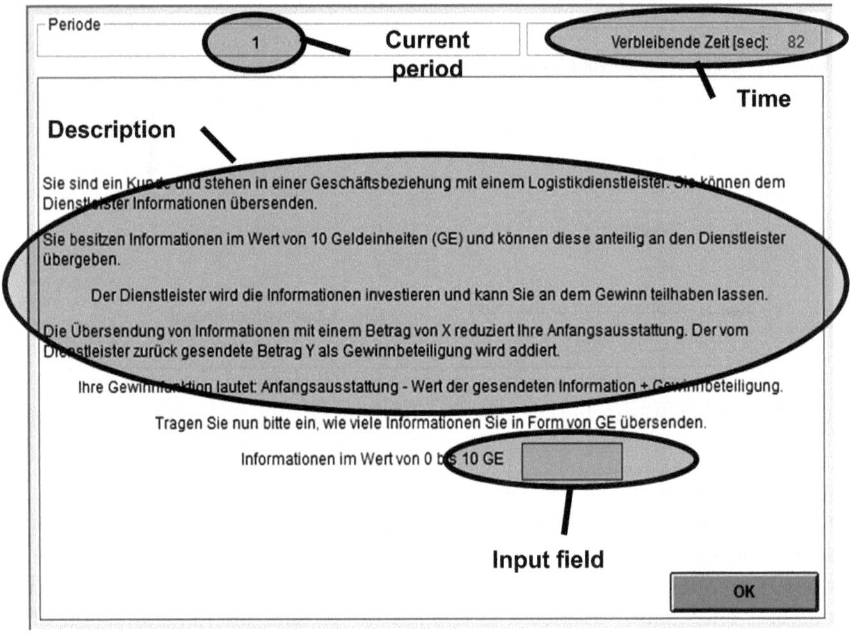

Fig. 3 On-screen interface for the participants (sender)

Payment

The payments of the participants depend on a performance-based and a fixed amount, a so-called show-up fee. The sender earns 10 MU minus the amount sent plus the amount returned by the responder. The responder earns 5 MU plus the amount that the sender has sent, multiplied by 3, minus the amount returned. The participants have the opportunity to earn an average of 8 € for about 45 min, including 2 € show-up fee.

Experimental Results

The presented results in this chapter are based on the main experiments. Table 1 shows a summary of the main descriptive results of the experiment.

The average quantity sent by the sender amounts to 5.14 MU in the control treatment and 6.88 MU in the 20 %-treatment. The senders hence send on average 25 % more with the guaranteed revenue share. The amount received by the senders is higher in the revenue-share setting as in the control treatment. While the responders send back a moderate higher amount (13 %) in addition to the revenue share,

Table 1 Average amounts exchanged and earned

	Control treatment	20 %-Treatment
Average amount sent by sender (MU)	5.14	6.88
Average amount sent by responder (MU)	6.67	7.58
Average amount sent by responder plus revenue share	6.67	11.24
Average profit sender (MU)	11.53	14.82
Average profit responder (MU)	13.76	13.92

this surpasses considerable the amount (40 %) in the control setting. These remarks are illustrated by the diagrams in Fig. 4.

The diagrams show the different investments and profits of the sender and responder sorted by the initial investment of the sender. The bar charts show that in the control treatment the responders send in 16 of the 20 groups a higher amount back than the initial investment. In the 20 %-treatment, the initial investments of six senders are ten (max. amount). Furthermore, in 11 cases the responder sends a higher amount back as the sender has sent to them, but with the revenue share of 20 % the amount rises by 27 %. Two responders send back nothing.

To prove the hypothesis (H): "With the guaranteed revenue share the overall profit will increase compared to the control setting", we compared the applied 20 %-treatment with the result of the control setting. The average group profit of the control treatment is 24.52 MU and the one in the 20 %-treatment is 28.76 MU. There is an increase of 14 %. To test the significance of the increase, the Mann–Whitney-U (MWU) non-parametric statistical test can be used for related samples if the control treatment and 20 %-treatment are being compared (Anderhub et al. 2002). The test results show that there is a significant difference in the total profit of the pairs ($p_{\text{total profit}} = 0.011$, one sided, $\alpha = 0.05$, $N = 20$). Based on $\alpha = 0.05 > p_{\text{total profit}}$, the null hypothesis that the profit of the responder is the same in both treatments can be rejected, so the above-mentioned hypothesis is supported. This result implies that the guaranteed revenue share of 20 % increases the overall profit of the pairs. Therefore, the mutual profit benefits from guaranteed revenue share.

If we look how the profit is allocated between the participants, we can see that the profit of the sender is significantly different between the treatments ($p_S = 0.005$, one sided, $\alpha = 0.05$, $N = 20$). However, in contrast to this, the profit of the responder is not significantly different between the treatments ($p_R = 0.333$, one sided, $\alpha = 0.05$, $N = 20$). Therefore, the null hypothesis must be maintained. Since the overall profit is solely dependent on the initial investment of the sender, the allocation of the profits is extremely interesting. As shown in Table 1, the profit of the sender increases disproportionally compared to the profit of the responder. This implies the sender profits more from the revenue-sharing contract in this experiment.

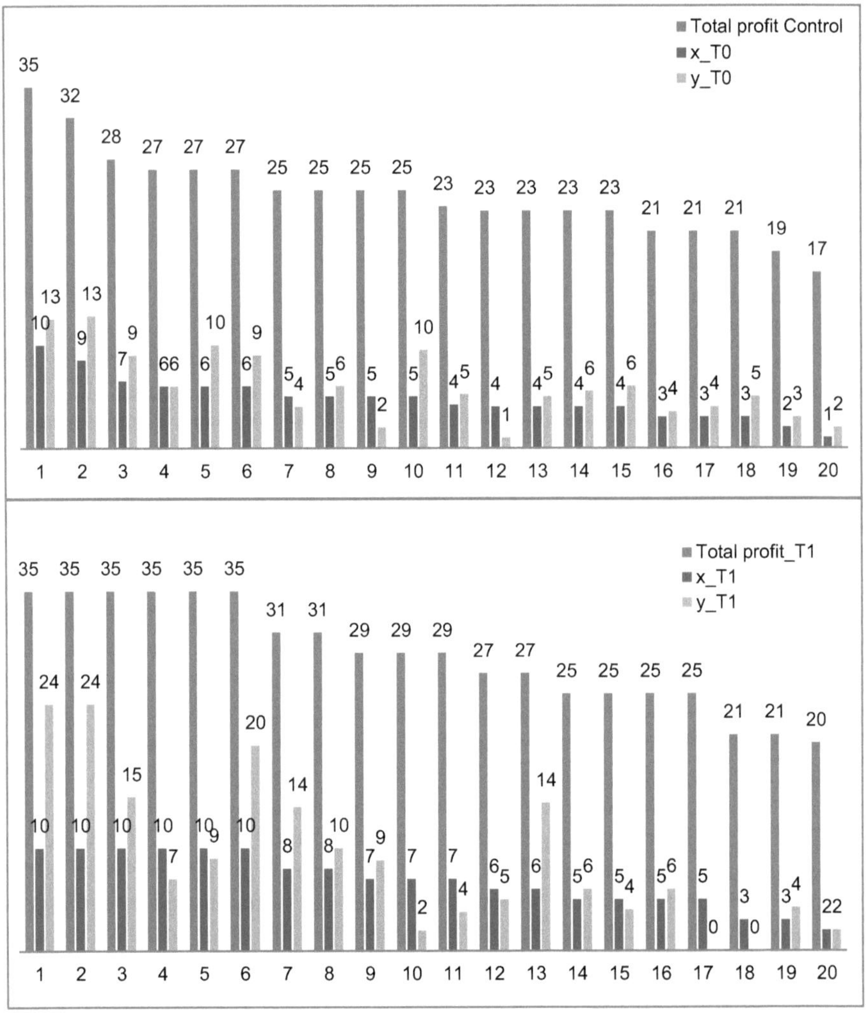

Fig. 4 Payments and exchanged amounts in Money Units (MU) sorted by pairs

Conclusion

To embrace the research question whether the revenue-sharing contract encourages the information disclosure, the laboratory experiments give first implications. The results can be interpreted in different scenarios like for example in the relationship of production-logistics, logistics-traffic or production-traffic. Our results of the laboratory experiment show that the sender (hereafter for example production in business relationship with logistics) sends a higher amount in the revenue-sharing (20 %-treatment) setting and additionally is rewarded by the responder (logistics) for the investment in the relationship. By a higher investment of the sender, the responder is determined to send back a higher amount including the revenue share.

The main findings and insights from the experiment can be summarized as follows: The first insight is that production benefits by sharing information to logistics. However, it has been shown that the guaranteed revenue share of 20 % increases the overall profit of the pairs. Therefore, both production and logistics benefit from a guaranteed revenue share, but production gains more from the investment in the business relationship than logistics. Logistics will not benefit equally from the shared data. This means that information sharing can contribute to be used as an effective means to improve supply chain efficiency.

The second finding is that the results implicate that revenue sharing can serve as an incentive or pricing component to get more information from production and provide efficient and effective services for production. It has been found that production is willing to provide more information, if it gets a guaranteed revenue share. In our experiment, a 25 % increase of the sent amount of production is observed. Due to the higher information disclosure of production, logistics should be able to plan their process more efficiently and effectively. This eventually enables logistics to offer a better service for a lower price.

The above findings on the impact of revenue sharing in business relationships provide useful insight into production's willingness to provide internal information to their logisticians.

There are limitations due to the validation and transfer of the result to practice. Especially, the kind and quality of the disclosed information has to be clarified. Furthermore, the calculation and verification of the profit generated by logistics must be determined and the appropriate kind of allocation to enhance information provision needs to be tested systematically. The experiment could be played with different revenue-sharing levels to analyse the actor's behaviour in different allocation scenarios.

With these results, we have demonstrated an incipient stage of research on how logistics is able to get more information from production by using a guaranteed revenue share as incentive. Based on revenue sharing in the business relationship, the information asymmetry between production, logistics and traffic can be reduced. Revenue sharing as incentive to get information for the design and execution of the processes in the business relationship seems to be a suitable mean. To ensure the practical relevance for a logistician, we plan to validate the results with experts from practice. Further laboratory experiments with managers of the different fields could be very interesting to gain a better insight of the application of revenue sharing in practice.

References

Anderhub, V., Engelmann, D., & Güth, W. (2002). An experimental study of the repeated trust game with incomplete information. *Journal of Economic Behavior and Organization, 48*(2), 197–216.

Anderson, J. C., & Thomson, J. B. (2000). Pursuing risk-sharing, gain-sharing arrangements. *Marketing Management, 9*(2), 40–47.

Bazerman, M. H. (2002). *Judgment in managerial decision making*. New York, NY: Wiley.

Bellantuono, N., Giannoccaro, I., Pontrandolfo, P., & Tang, C. S. (2009). The implications of joint adoption of revenue sharing and advance booking discount programs. *International Journal of Production Economics, 121*(2), 383–394.

Berg, J., Dickhaut, J., & McCabe, K. (1995). Trust, reciprocity, and social history. *Games and Economic Behavior, 10*, 122–145.

Boero, R., Bravo, G., Castellani, M., & Squazzoni, F. (2009). Reputational cues in repeated trust games. *The Journal of Socio-Economics, 38*(6), 871–877.

Bradley, P. (1995). Third parties gain slow, cautious buyer support. *Purchasing, 18*, 51–52.

Cachon, G. P., & Lariviere, M. A. (2001). Turning the supply chain into a revenue chain. *Harvard Business Review, 79*(3), 20–21.

Cachon, G. P., & Lariviere, M. A. (2005). Supply chain coordination with revenue-sharing contracts: Strengths and limitations. *Management Science, 51*(1), 30–44.

Chakravarty, A. K., & Werner, A. S. (2011). Telecom service provider portal revenue sharing and outsourcing. *European Journal of Operational Research, 215*(1), 289–300.

Coricelli, G., Morales, L. G., & Mahlstedt, A. (2006). The investment game with asymmetric information. *Metroeconomica, 57*(1), 13–30.

Dana, J., & Spier, K. E. (2001). Revenue sharing and vertical control in the video rental industry. *The Journal of Industrial Economics, 44*(3), 223–245.

Daugherty, P. J., Myers, M. B., & Richey, R. G. (2002). Information support for reverse logistics: The influence of relationship commitment. *Journal of Business Logistics, 23*(1), 85–106.

Deck, C., & Smith, V. (2013). Using laboratory experiments in logistics and supply chain research. *Journal of Business Logistics, 34*(1), 6–14.

Fahr, R., & Irlenbusch, B. (2000). Fairness as a constraint on trust in reciprocity—Earned property rights in a reciprocal exchange experiment. *Economics Letters, 66*, 275–282.

Fischbacher, U. (2007). z-Tree. Zurich toolbox for readymade economic experiments. *Experimental Economics, 10*, 171–178.

Füllbrunn, S., Richwien, K., & Sadrieh, A. (2011). Trust and trustworthiness in anonymous virtual worlds. *Journal of Media Economics, 24*(1), 48–63.

Giannoccaro, I., & Pontrandolfo, P. (2004). Supply chain coordination by revenue sharing contracts. *International Journal of Production Economics, 89*(2), 131–139.

Giannoccaro, I., & Pontrandolfo, P. (2009). Negotiation of the revenue sharing contract: An agent-based systems approach. *International Journal of Production Economics, 122*(2), 558–566.

Highhouse, S. (2009). Designing experiments that generalize. *Organizational Research Methods, 12*(3), 554–566.

Houser, D., Schunk, D., & Winter, J. (2010). Distinguishing trust from risk: An anatomy of the investment game. *Journal of Economic Behavior and Organization, 74*(1–2), 72–81.

Katok, E. (2011). Using laboratory experiments to build better operations management models. *Technology, Information and Operations Management, 5*(1), 1–86.

Klug, F. (2010). *Distributionslogistik im Automobilbau. Logistikmanagement in der Automobilindustrie* (pp. 427–445). Berlin: Springer.

Knemeyer, A. M., & Murphy, P. R. (2004). Evaluating the performance of third-party logistics arrangements: A relationship marketing perspective. *The Journal of Supply Chain Management, 40*(Winter), 35–51.

Knemeyer, A. M., & Naylor, R. W. (2011). Using behavioral experiments to expand our horizons and deepen our understanding of logistics and supply chain decision making. *Journal of Business Logistics, 32*(4), 296–302.

Kumar, M., Vrat, P., & Shankar, R. (2006). A multi-objective 3PL allocation problem for fish distribution. *International Journal of Physical Distribution and Logistics Management, 36*(9), 702–715.

Lakhal, S. Y. (2006). An operational profit sharing and transfer pricing model for network-manufacturing companies. *European Journal of Operational Research, 175*(1), 543–565.

Lambert, D. M., Emmelhainz, M. A., & Gardner, J. T. (1999). Building successful logistics partnerships. *Journal of Business Logistics, 20*(1), 165–181.

Langley. (2009). Third-party logistics study: The state of logistics outsourcing. Results and findings of the 14th annual study. Accessed October 14, 2014, from www.3plstudy.com

Langley. (2012). Third-party logistics study: The state of logistics outsourcing. Results and findings of the 16th annual study. Accessed July 28, 2014, from www.3plstudy.com

Lewicki, R. J., Tomlinson, E. C., & Gillespie, N. (2006). Models of interpersonal trust development: Theoretical approaches, empirical evidence, and future directions. *Journal of Management, 32*, 991–1022.

Liu, W., Xu, X., & Kouhpaenejad, A. (2013). Deterministic approach to the fairest revenue-sharing coefficient in logistics service supply chain under the stochastic demand condition. *Computers and Industrial Engineering, 66*, 41–52.

Malhotra, D. (2004). Trust and reciprocity decisions: The differing perspectives of trustors and trusted parties. *Organizational Behavior and Human Decision Processes, 94*, 61–73.

Malhotra, D., & Murnighan, J. K. (2002). The effects of formal and informal contracts on interpersonal trust. *Administrative Science Quarterly, 47*, 534–559.

Marasco, A. (2008). Third-party logistics: A literature review. *International Journal of Production Economics, 113*(1), 127–147.

Mohr, J., & Spekman, R. (1994). Characteristics of partnership success: Partnership attributes, communication behavior, and conflict resolution techniques. *Strategic Management Journal, 15*(2), 135–152.

Mortensen, O., & Lemoine, O. W. (2008). Integration between manufacturers and third party logistics providers? *International Journal of Operations and Production Management, 28*(4), 331–359.

Nahapiet, J., & Ghoshal, S. (1998). Social capital, intellectual capital, and the organizational advantage. *Academy of Management Review, 23*(2), 242–266.

Narayanan, V. G., & Raman, A. (2004, November). Aligning incentives in supply chains. *Harvard Business Review*, 94–102.

Pfohl, H.-C. (2010). *Logistiksysteme. Betriebswirtschaftliche Grundlagen* (8th ed.). Berlin: Springer.

Pillutla, M., Malhotra, D., & Murnighan, J. K. (2003). Attributions of trust and the calculus of reciprocity. *Journal of Experimental Social Psychology, 39*, 448–455.

Razzaque, M. A., & Sheng, C. C. (1998). Outsourcing of logistics functions: A literature survey. *International Journal of Physical Distribution and Logistics Management, 28*(2), 89–107.

Rigdon, M. (2009). Trust and reciprocity in incentive contracting. *Journal of Economic Behavior and Organization, 70*(1–2), 93–105.

Selviaridis, K., & Spring, M. (2007). Third-party logistics: A literature review and research agenda. *International Journal of Logistics Management, 18*(1), 125–150.

Siemsen, E. (2011). The usefulness of behavioral laboratory experiments in supply chain management. *The Journal of Supply Chain Management, 47*(3), 17–18.

Stank, T. P., Daugherty, P. J., & Ellinger, A. E. (1996). Information exchange, responsiveness and logistics provider performance. *The International Journal of Logistics Management, 7*(2), 43–57.

Tokar, T. (2010). Behavioral research in logistics and supply chain management. *The International Journal of Logistics Management, 21*(1), 89–103.

Tsai, M.-C., Lai, K., Lloyd, A. E., & Lin, H.-J. (2012). The dark side of logistics outsourcing—Unraveling the potential risks leading to failed relationships. *Transportation Research Part E, 48*, 178–189.
van der Rhee, B., van der Veen, J. A. A., Venugopala, V., & Nalla, V. R. (2010). A new revenue sharing mechanism for coordinating multi-echelon supply chains. *Operations Research Letters, 38*, 296–301.
Wallenburg, C. M., & Wieland, A. (2010). Widerstandsfähige Logistiknetze durch robuste und agile Supply-Chain-Strategien. In H. Wolf-Kluthausen (Ed.), *Jahrbuch Logistik 2010* (pp. 52–57). Korschenbroich.
Welbourne, T. M., & Mejia Gomez, L. R. (1995). Gainsharing—A critical review and a future research agenda. *Journal of Management, 21*(3), 559–609.
Wilding, R., & Juriado, R. (2004). Customer perceptions on logistics outsourcing in the European consumer goods industry. *International Journal of Physical Distribution and Logistics Management, 34*(8), 628–644.
Yao, Z., Leung, S. C. H., & Lai, K. (2008a). The effectiveness of revenue-sharing contract to coordinate the price-setting newsvendor products' supply chain. *Supply Chain Management: An International Journal, 13*(4), 263–271.
Yao, Z., Leung, S. C., & Lai, K. (2008b). Manufacturer's revenue-sharing contract and retail competition. *European Journal of Operational Research, 186*(2), 637–651.
Zhang, Q., Vonderembse, M. A., & Lim, J. S. (2005). Logistics flexibility and its impact on customer satisfaction. *International Journal of Logistics Management, 16*(1), 71–95.

Supply Chain Risk Management in the Automotive Industry: Cross-Functional and Multi-tier Perspectives

Liyuan Wang, Kai Foerstl, and Friso Zimmermann

Introduction

The integration of different functional areas, such as supply management, production, logistics, and transportation, within one entity and across multiple supply chain tiers, remains a major challenge. This study focuses particularly on such cross-functional integration processes of the multi-tier supply chain risk management in the automotive industry. The objective of the study is to provide tools for managers to identify, assess, and mitigate risks across their multi-tier supply chains. This holistic perspective on supply chain risk management reflects the findings of Dynamo PLV—an interdisciplinary research project. This research project incorporated supply chain risk management as a cross-functional practice aiming at fuelling supply chain responsiveness and cost optimisation.

A large number of risk management examples demonstrate the relevance of this topic. In 2013, a major Japanese automotive Original Equipment Manufacturer (OEM) recalled nearly four million vehicles worldwide due to safety concerns caused by quality problems originating from their airbag supplier. Similar recalls are almost never caused by the manufacturer's own engineering lapses, but by a deviation in some material or process from a supplier, sub-supplier or sub-sub-supplier (Financial Times 2013). According to the third annual Allianz Risk Barometer, which surveyed over 400 corporate insurance experts from more than

L. Wang (✉) • F. Zimmermann
EBS Universität für Wirtschaft und Recht, EBS Business School, Institute for Supply Chain Management, Burgstr. 5, 65375 Oestrich-Winkel, Germany
e-mail: liyuan.wang@students.ebs.edu; friso.zimmermann@students.ebs.edu

K. Foerstl
Professor für Supply Chain Management und Logistik, German Graduate School of Management & Law (GGS), Bildungscampus 2, 74076 Heilbronn, Germany
e-mail: kai.foerstl@ggs.de

E. Abele et al. (eds.), *Dynamic and Seamless Integration of Production, Logistics and Traffic*, DOI 10.1007/978-3-319-41097-5_7

30 countries, supply chain risk is ranked among the highest in the 2014 Risk Barometer. Within the context of globalisation of markets (Levitt 1983), the biggest risks to a company are not restricted to disruption or glitches within the organisation. Instead, a company's ability to perform depends strongly on its supply chain partners. Thus, firms should not only seek opportunities to enhance competitiveness through internal efficiency improvements, but also focus on their upstream supply chain risks and opportunities (Kajüter 2003).

Supply chain perils can damage supply chain infrastructure and significantly impact the private sector's financial performance. As Waters (2007) states, "an interruption to the supply chain can have widespread effects. Shareholder return typically falls by 7–8 % on the day that a disruption is announced; operating income falls by 42 % and return on assets is down by 35 %." Based on a sample of 519 announcements made during 1989–2000, Hendricks and Singhal (2003) found that glitch announcements decreased shareholder value by about 11 %. After all, it does not matter where in the supply chain a risk or disruption is caused; what matters is the detriments resulting from this risk or disruption.

Business leaders, who can exploit the complexity of the entire supply chain, deliver value and drive business sustainability for both customers and trading partners are increasingly being hunted by boards of directors. According to Gartner's report in 2011, the number of supply chain executives elected as or reporting directly to the Chief Executive Officer (CEO) has increased from 30 % in 2005 to 68 % in 2010. Thus, the need for agility and a better understanding of the supply network takes precedence in the C-suite of companies (Winter 2012).

Our motivation in studying cross-functional, multi-tier supply chain risk management in the automotive industry is driven by two factors, namely the observed research gap in academia and the current business needs in the knowledge exchange on this topic. The reason for focusing on the automotive industry is that as companies in the industry become increasingly global in their supply, production and sales footprints, this vulnerability is increasingly revealed. In particular, supply problems often arise from port disturbances, labour strikes, fuel price fluctuations and the general volatility that characterises emerging markets. Risk or disruptions in the product flow may have devastatingly negative consequences both for an individual firm and for the supply chain (Paulsson 2007).

The guiding research question in our study is whether and how companies in the automotive industry deal with risks in their multi-tier supply chains. The purpose of this study is to analyse the status quo of cross-functional, multi-tier supply chain risk management in the automotive industry in order to advance knowledge and to provide a cross-company learning opportunity for practitioners. We shall do this by comparing and contrasting supply chain risk management practices across multiple functional and tier levels covering OEM, first-tier supplier(s), and second-tier supplier(s).

Globalisation, specialisation (focusing on core competencies) and lean production provide the automotive industry with numerous opportunities and at the same time many challenges. Companies at different tier levels in the industry thrive by achieving and sustaining high levels of customer satisfaction and cost advantages.

Applying this dual approach to attaining both cost efficiency and responsiveness makes supply chains more vulnerable to risks such as delays in transportation or production schedules, quality problems from suppliers or force majeure. Hence, effective management of these risks seems to demand an integrated approach towards supply chain information and material flows.

Questions raised in this study include:

- What kind of contemporary supply chain risks exist in the automotive industry?
- Which tools do, or should, companies use to identify, assess and mitigate risks in their multi-tier supply chains?
- What is the importance of a holistic perspective for multi-tier supply chain risk management?
- What are the cross-learning effects among various firms along the automotive supply chain?

Although for some companies, successfully managing the multi-tier supply chains and the inherent risks can offer a real competitive advantage, for other companies, multi-tier supply chain risk management has become a prerequisite for success or even survival. At the same time, there exists neither a multipurpose solution nor a quick fix for initiating the best supply chain risk management practice. Only a context-specific and suitable supply chain risk management approach that is aligned with the overarching corporate performance goal while at the same time mitigating supply chain-related risks allows companies a sustained protection from the potentially most detrimental impacts on their operations.

Theoretical Background and Research Methodology

Basics and Definitions

Supply chains are composed of entwined sets of interacting entities of nearly unimaginable complexity and sophistication. Because of this intrinsic nature, many argue that the term "supply chain" paints rather a simple picture and should therefore be replaced by supply networks or supply and demand networks. Alternative terms refer to a "logistics channel" to emphasise marketing, a "process" to emphasise operations and a "value chain" to emphasise value added (Porter 1985). However, such differences refer simply to semantics rather than to content.

In the absence of a commonly agreed upon definition of the term "supply chain", in this study, the supply chain is defined as (a) the processes from the supply of initial raw materials to the ultimate consumption of the finished product linking across supplier-user companies and (b) functions within and outside a company that enable the value chain to make products and provide services to the customer.

According to the Council of Supply Chain Management Professionals (CSCMP), "(*supply chain management*) encompasses the planning and management of all activities involved in sourcing and procurement, conversion and all logistics management activities." The concept also places emphasis on coordination and collaboration between channel partners, which refers to suppliers, intermediaries, third-party service providers and customers. Guided by this definition, we started our analysis from the position of OEMs and covered their upstream channels including first-tier and second-tier suppliers.

Risk itself can be described as disruption, vulnerability, uncertainty, disaster, peril and hazard (Ghadge et al. 2012). *Supply chain risk* can be broadly defined as exposure to an event which causes disruption, thus affecting the efficient management of the supply chain network.

A diverse classification of supply chain risks is available in the literature. Based on the work of Norrman and Jansson (2004), we divided supply chain risk content into six categories, namely (1) market risk, (2) liquidity risk, (3) volume/capacity/demand risk, (4) counterparty risk, (5) operating risk and (6) risk interrelations.

1. *Market risk* refers to the possibility of losses caused by adverse movements in market prices. Market risk is derived from sources such as fluctuations in exchange rates, interest rates and prices of commodities or sourced components.
2. *Liquidity risk* refers to the inability to meet obligations derived from supply chain activities (e.g. payments to suppliers).
3. *Volume/capacity/demand risk* derives from fluctuations in production and demand levels such as those caused by changes in customer preferences.
4. *Counterparty risk* arises from the possible non-performance of contractual obligations by counterparties, in particular suppliers or customers.
5. *Operating risk* denotes the risk of loss resulting from inadequate or failed internal processes, people and systems, or from external events. Examples include inadequate procurement forecasts, production breakdowns and workers' strikes.
6. *Risk interrelation* refers to interactions between different risks. For instance, outsourcing can mitigate operational risk in an organisation while control over the outsourced processes decreases, and thus, counterparty risk increases. In many cases, risk interrelations can also lead to an amplification of the underlying risks and their potential impacts. As an example, consider a company using EUR as its reporting currency that plans to purchase components against the USD. In this context, the impact of rising component prices in USD can be aggravated by an appreciation of USD against EUR.

After viewing different notions of *supply chain risk management*, we define the term in this study as the management of supply chain risk through coordination or collaboration among the supply chain partners so as to ensure profitability and continuity.

Supply chain risk management strategies are classified in *proactive* and *reactive* approaches. Proactive risk management refers to the approach that avoids potential risk events from occurring.

Reactive risk management is to respond to risk events that have already occurred so that consequences can be addressed. One of the reactive risk management strategies that many companies follow is contingency planning. This type of risk response, where there is a defined process or response plan, will limit the impact or duration of the risk event after a risk has occurred. Different key proactive and reactive risk mitigating strategies are discussed in the literature (Ghadge et al. 2012).

Proactive Risk Management Strategies are:

- Supplier development/management: risk sharing through contract manufacturing, contractual governance, dual/multi-sourcing
- Supply chain contracts: developing incentive contracts, mix and volume flexibility contracts for risk mutual benefits, buffer stock
- Product/process management: product variety, postponement, product design and delivery management
- Supplier relationship: supplier collaboration through improved confidence, cultural adaptation, continuous coordination

Reactive Risk Management Strategies are:

- Contingency planning: strategic event management plan, enhanced flexibility in options
- Disaster management: robust recovery, rebuilding of supply chain, resource utilisation/management, scenario analysis for future disruptions
- Demand management: operational rerouting, shifting customer demand, dynamic pricing

Research Methodology

At first, a literature review was performed via a thorough evaluation of existing research into risk and risk management in other disciplines, such as finance, insurance and engineering, and also in our focal areas of supply chain management and supply chain risk management.

Our motivation for adopting the empirical qualitative research approach was driven by "a shortage of empirical research in the area of supply chain risk management" (Sodhi et al. 2012), and limited empirical research into the risks prevalent in supply chain networks (Harland et al. 2003).

In addition, more qualitative research is needed to better understand the variety of different risks and the processes that allow active prevention and management (Kern et al. 2012). Research methods suitable for capturing the holistic as well as the dynamic behaviour of risk within supply chain networks were found to be clearly lacking (Ghadge et al. 2012). Thus, a multiple case study involving an analysis of not only a firm's direct suppliers (first-tier suppliers) but also its suppliers' suppliers (second-tier suppliers) was conducted. This seemed necessary in order to attain a more holistic picture of the supply chain risk management

Table 1 Coverage and distribution of the interviews

Tier industry	OEM	First-tier supplier	Second-tier supplier
Automotive Industry	6 Interviews	5 Interviews	3 Interviews

approaches practiced across the multiple entities in a supply chain (Trkman and McCormack 2009).

We applied a qualitative research approach based on case studies, which included in-depth semi-structured interviews with 14 managers from procurement and logistics, sales and risk management functions. Additionally, we reviewed empirical data such as internal documents regarding the key elements and processes of supply chain risk management in their respective firms.

A case study is a research strategy which concentrates on perceiving the dynamics present within single settings (Eisenhardt 1989). Throughout the year 2013, we conducted in-depth interviews with representatives from the automotive industry covering OEMs, first-tier suppliers and second-tier suppliers.

We used semi-structured interviews as our primary data collection method to systematically record supply chain risks prevalent in the supply chains and the processes associated with managing these risks (Romeike and Hager 2009). After transcribing the interviews, we coded them according to scientific qualitative data analysis techniques (Strauss and Corbin 1989). The coding method provided us with a thematic analysis approach and allowed us to compare and contrast risks and management practices based on their strengths and intensities.

The structured interview guideline contained 15 questions. These questions were designed to shed light on the following aspects: What are the supply chain risks in the studied companies? What are the similarities and differences? What are the sources and drivers of supply chain risks? In terms of proactivity and/or reactiveness, what is the risk management strategy of the interviewed company? What does the supply chain risk management process look like? How do the interviewed companies integrate multi-tier suppliers in order to manage supply chain risks?

Based on the interview results and other research outcomes, we provided further suggestions and recommendations. Contributors from eight companies participated in our 2-h interviews, held in English. The coverage and distribution of the interviews are illustrated in Table 1.

Nearly half of our interviewees have industry experience of more than 10 years. Seventy per cent of them hold middle or top management positions in the corresponding division.

Data Analysis

Major Findings

We found strong evidence to support multi-tier supply chain risk management as opposed to single-tier supply chain risk management. Multi-tier visibility and collaboration enable manufacturers to synchronise planning and to coordinate the execution of activities based on information from all relevant direct partners and other stakeholders along the chain. We found that such an approach provides opportunities to efficiently manage supply chain risks and to reduce the impact and likelihood of disruption. Nevertheless, we identified some gaps in corporate practice when it was evaluated according to acclaimed holistic Supply Chain Risk Management approaches. The gaps were as follows:

Gap 1: Limited Focus on the Multi-tier Levels in the Supply Chain: Supplier risk management is often simplified as direct supplier risk management, which neglects the risk interconnectivity and interaction across the chain and the supply network.

Gap 2: Quantitative Risk Evaluation: Many companies relied on a qualitative view of risk evaluation without real quantifiable evidence and potential to aggregate risks across dimensions or supply chain stages.

Gap 3: Supply Chain Risk Management as a Dynamic Process: Re-evaluation of the residual risk was not fully integrated in the supply chain risk management process.

Based on the gaps found in our research, we introduced four approaches to assist companies with mitigating multi-tier supply chain risks and seizing potential business opportunities.

1. The Three-Circle Model helps managers to identify risks from a multi-tier supply partner network. This model can be applied in all companies along the supply chain.
2. The Risk Portrait of the Multi-tier Supply Chain Partner Approach operationalises the Three-Circle Model to increase the risk visibility and comparability of the selected multi-tier suppliers.
3. The quantitative approach to assess multi-tier supply chain risk includes five basic steps, namely prioritisation of supply chain risk by using tools such as Risk Mapping, quantification of the risk impact, modelling joint risk factor dynamics, calculation of the risk probability distribution and aggregation of supply chain risks. It allows the hard quantification of the previously visualised risks.
4. The Residual Risk Diagram treats the risk management process as a dynamic system. This model concentrates on the recurring re-evaluation of the residual risk of within multi-tier supply chains.

In Table 2, we summarise and group the interview results into four areas of supply risk management which we queried based on our Interview Guideline: (1) source of supply chain risk, (2) risk management strategy (proactive vs. reactive), (3) risk management coverage and depth along the supply chain and (4) risk management process.

Table 2 Main interview findings in the automotive industry

Automotive industry			
Unit of analysis	OEMs	First-tier suppliers	Second-tier suppliers
Source of supply chain risk	• Demand and operating risks due to lean management and outsourcing • Counterparty and regulation risks • Risk of supply chain sustainability	• Operating risk due to on-time delivery and lead time • Market risks especially due to raw material price • Supply chain complexity • Counterparty risk of the suppliers	• Market and demand risks • Operating risk • Counterparty risk of the B2B customer
Risk management strategy: proactive vs. reactive	• Proactive and reactive, such as early supplier involvement and supplier collaboration • Long-term perspective on supplier evaluation • Supplier relationship building strategy	• Both proactive and reactive, such as supplier development and contingency planning • Dual sourcing strategy	• Both proactive and reactive • Supply chain cluster management • Supplier development through contract manufacturing
Risk management coverage and depth along the supply chain	• Direct coverage only on selected first tier • Downstream to car dealer • Competitive and contractual relationship with first-tier suppliers	• No risk transfer but risk sharing (such as the fluctuation of raw material price) with OEMs • Only first tier but with a certain impact on and indirect monitoring of sub-tier	• Upstream only first tier • Monitoring of credit standing of key account customers
Risk management process	• Early warning indication (quality and delivery) • Regular evaluation of key suppliers • Lack of sufficient re-evaluation of the residual risk	• Early warning indication (quality and delivery) • Regular evaluation of key suppliers • Long-term agreements with raw material suppliers • Lack of sufficient re-evaluation of the residual risk	• Risk management process at both corporate and division levels • Early warning indication

Supply Chain Risk Categories

In the Definitions section, we divided the supply chain risk into six types. Our interview findings showed that the top three risks that absorb the most corporate resources and demand the most managerial involvement in the automotive industry are market risk, operating risk, and volume/capacity/demand risk.

To be more specific, market risk—the risk of losses in situations arising from movements in market prices—caused the most concern within our interviewees. We also found that internal risks, especially business and operational risks, such as

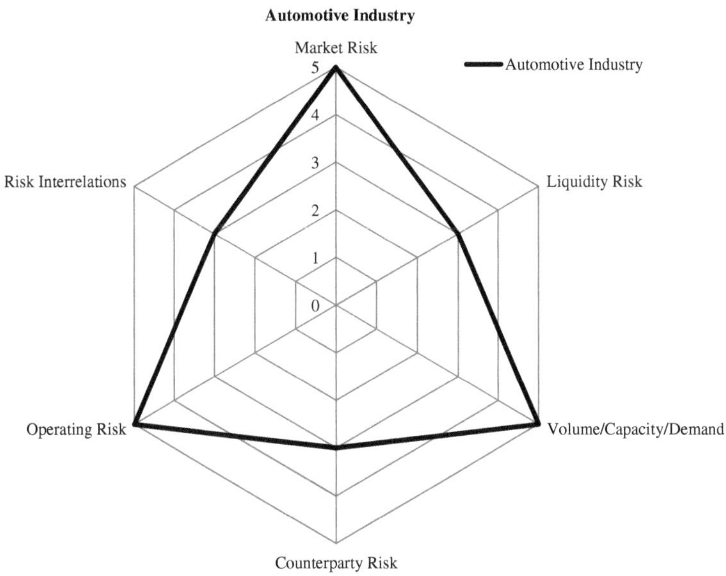

Fig. 1 Industry comparison: supply chain risk profiles

raw material supplier failure, product quality failure, finished product manufacturing failure and the inability of counterparties to meet contractual obligations, rank next in terms of causing concern to those interviewed companies.

The interviewees' perspectives of the emphasis placed in their organisations on different types of supply chain risk are shown in Fig. 1. The 1–5 scale measures the extent of the interviewee's involvement in the corresponding risk category, in which 1 represents no involvement and 5 the highest degree of involvement.

In addition, we found that risk events with straightforward and tangible consequences seem to attract more attention from managers than those with severe, but hidden or non-quantifiable, impacts. We believe there are two major explanations here. Firstly, managers are likely to have experienced these risks several times in their careers and thus feel empowered to mitigate them. This is the "risk comfort zone", a set of frequently occurring, more preventable routine business risks. As a result of the frequent occurrence of such risks, organisations are equipped to analyse and mitigate them. Secondly, the performance appraisal system for the manager may also have a direct impact on the manager's risk priority setting. As one of the informants said, managers would hardly be given credit for managing risks that didn't happen or have invisible impacts. Managers are driven to fulfil their work missions by solving the most urgent and important supply chain risks, and thus there is less capacity or resource availability for investing in safeguarding against those hidden risks or risks with non-quantifiable impacts.

Drivers of Supply Chain Risk

As stated in the previous section, market risks and internal risks, especially business and operational risks, are the most significant risk sources for the automotive industry. In our interviews, we discovered that among a number of factors that contribute to an increase in the risk of a supply chain disruption, the most decisive source or driver is the global nature of today's supply chains. The evolving global nature of supply chains and business models, such as just-in-time, lean management, single-sourcing, outsourcing and offshoring, has led to changes in the distributed areas of supply chain risks, which is the common feature being researched here.

In response to the trend of globalisation and global competition, our interviewed companies were found to have undertaken a variety of business practices to improve cost efficiency. Since the 1990s, focusing on the supply chain has brought about improvements in cost efficiency (Christopher and Lee 2004). Leaping ahead, companies started to implement concepts streamlining supply chain processes (Childerhouse and Towill 2003), such as just-in-time, just-in-sequence, virtual inventory and reducing the number of distribution facilities in order to create lean supply chains (Svensson 2004; Thun et al. 2007). The lack of safety stock and reduced resilience in responding to sudden and unanticipated changes in demand become the first major drivers of supply chain risks faced by practitioners.

A second trend resulting from the globalisation of supply chains is outsourcing and offshoring. Outsourcing and offshoring increase the dependency of companies along the chain and the complexity of the whole supply chain network. Thus, it is hard for companies to gain a holistic and transparent picture of the entire supply chain network. The more complex a network is, the more interfaces there will be, and the greater the vulnerability (Peck 2005). For instance, in 2013 major Japanese automotive OEMs recalled nearly four million vehicles worldwide due to a safety problem with airbags supplied by the Takata Corp. (Financial Times 2013). According to our interview subjects from the automotive industry, similar car recalls are almost never caused by the manufacturer's own engineering lapses, but by a deviation in some material or process at a first-, second- or even end-tier supplier.

A third trend also stems from the globalisation of supply chains, which is characterised by longer shipping distances and the heightened complexity of trade. Companies have to account for factors such as currency shifts, political unrest, cultural differences, piracy and natural disasters. Due to the global nature of today's business, external events such as terrorist attacks, earthquakes and floods in one region can have ripple effects and thus jeopardise the whole supply chain. Such catastrophic events have reinforced the awareness of vulnerability along the entire supply chain for manufacturing firms (Knemeyer 2009), both downstream in dealing with customers and upstream in dealing with suppliers (Jüttner 2005). As reported by our interviewees, external environmental risk together with the growing market uncertainty of demand and the shrinking of product life cycles makes the job of matching supply with demand more complex than ever before. For instance, the 2010 volcanic eruptions in Iceland forced about 20 countries to close their airspace and the supply chains of our case study companies were unable to deliver parts to OEMs and end customers on time, which in turn caused certain negative impacts on the normal business operation and customer satisfaction.

Actual Supply Chain Risk Management Gaps and Potential for Improvement

As a result of insights gained from our interviews and the literature review, we identified three significant gaps between the practical realities of the business world and what is proclaimed as being the best practice in supply chain risk management in academic works. In the following section, we provide in-depth exploration of the three gaps.

Gap 1: Limited Focus on the Multi-tier Levels in the Supply Chain Supplier risk management is often simplified as direct first-tier supplier risk management, which neglects risk interaction and intersection with multi-tier suppliers.

Our interviewees agreed that supplier management practices adopted by OEMs have an impact on first-tier suppliers' performance, which then, in turn, impact on the OEM's quality and delivery performance to the customer. Similarly, the way in which first-tier suppliers communicate and manage second-tier suppliers shapes the performance of second-tier suppliers and ultimately influences how first-tier suppliers perform. In practice, the implementation of single-tier and/or multi-tier supply chain risk communication falls into two categories. These are illustrated in Fig. 2.

The left diagram in Fig. 2 represents about 50 % of the interviewees. In this model, the OEMs communicate only with first-tier suppliers. The flow of information and communication is indicated by lines in the diagram. Although there is information and communication exchange between first-tier suppliers and second-tier suppliers, a direct link between OEMs and second-tier suppliers is lacking. Although these automotive OEMs require their first-tier suppliers to regularly evaluate the second-tier ones as part of the performance evaluation of the first-tier suppliers, they normally do not actively engage in further sub-supplier management.

The other half of the respondents, represented by the right-hand diagram in Fig. 2, stressed that their OEMS do systematically monitor their second-tier suppliers via first-tier suppliers. These companies take a direct approach to second-tier supplier management and development by requiring suppliers to acquire goods and services from specific lower-tier suppliers and under specifically defined conditions. Companies sometimes also work directly with second-tier supplier alliance partners in solving problems through certification programmes and implementing alliance development strategies. The information and communication flow between

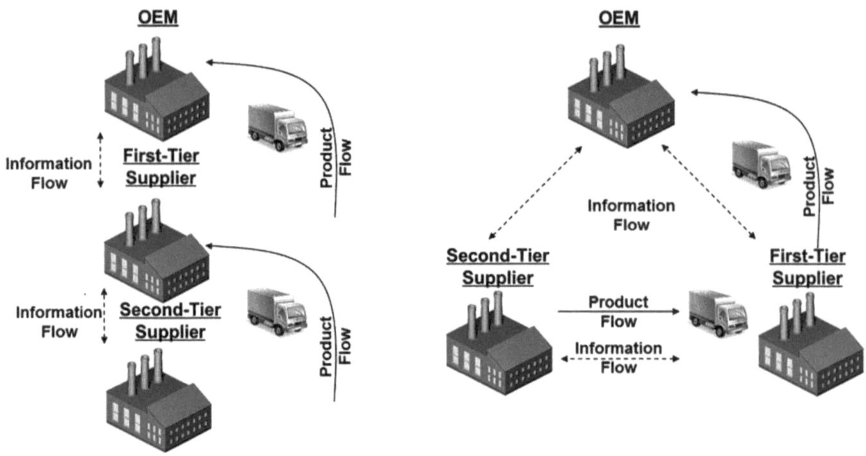

Fig. 2 Tier system of supply chain risk communication and product flow

OEMs and second-tier suppliers is sometimes channelled via first-tier suppliers, as indicated by the green dashed lines.

We concur with Choi and colleagues (2001) who state that those large organisations with sophisticated supply chain management practices do not have visibility over their own supply chain. As one of the company representatives explained in the interview, "It is not our job to monitor the risk of our second-tier suppliers, so we leave it to our first-tier suppliers to manage their lower-tier suppliers. And we are going to stick to this." In the end, the fundamental limiting factors for such practices are the significant amount of resources required and the executive commitment (Becks 2010).

As pointed out by Frohlich and Westbrook (2001), the arcs of supply chain risk management integration may extend to upstream and downstream supply chain partners. It is considered either as a narrow arc of integration with little integration beyond direct suppliers or a broad arc of integration with extensive integration of multi-tier supplier stages along the whole chain. However, one decisive premise of the implementation of the integration is that companies have to integrate internally first before they can engage in aligned supply chain risk management tasks with their external partners, such as multi-tier suppliers and business customers.

In our interviews, we found that some automotive OEMs have a broad arc of integration. However, in general there is a strong tendency to integrate and collaborate with only first-tier suppliers. Growing evidence indicates that the higher the level of integration and collaboration of multi-tier suppliers, the greater the potential benefits, both in terms of capturing the opportunities and mitigating the risks. There are three major benefits:

Firstly, broader integration and collaboration facilitate more seamless product launches. Close communication and coordination across multiple tiers of the supply base ensures that rapid production accelerates smoothly and cost effectively, especially in the product launch phase.

Secondly, direct multi-tier communication and information exchange helps to avoid or minimise the problems of asymmetric information and thus increases the velocity of information flow so that the quality of information used in the planning and operation processes will improve. This effectively reduces the risk of shortages, improves forecast consensus and reduces lead times, ultimately generating higher levels of service.

Thirdly, the overall performance of the supply base can be improved. In order to measure the performance of strategic suppliers, OEMs need to be connected to them, regardless of the tier. By tapping into supply and demand events, they can measure ongoing performance metrics, plus key metrics such as effectiveness in delivering against demand and flexibility in responding to changes. Lastly, multi-tier visibility creates a resilient and agile supply chain. With visibility into operational disruptions across the extended value chain, companies are able to see—and respond to—sources of supply chain risk before operations and supply chain performance are impacted (Becks 2010).

Gap 2: Quantitative Risk Evaluation Many companies relied on a qualitative view of risk evaluation without real quantifiable evidence.

As Taylor has already stated in 1999, "there is no doubt of the importance of quantitative models and computer based tools in decision making in today's business environment." As his work further states, this holds especially true for today's increasingly complex and global supply chains. According to Alvarenga (2012), anyone who claims to be managing supply chain risk without understanding subjects like real options, hedging, value at risk models, financial simulation and so on is more like a security guard than a real risk manager.

However, a common observation from our interviews was that managers are rarely trained in or practice quantitative concepts and methodologies for the evaluation and aggregation of risks across the supply chain. Thus, in these cases, decision-making can only be based on a qualitative view of supply chain risks. To underpin the significance of this issue, we wish to propose three major benefits that will assist companies in using quantitative methods in supply chain risk management.

Firstly, quantitative analysis can enhance the transparency of management decisions and provide accountability and objectivity to the policies of those in charge of supply chain risk management.

Secondly, the implementation of quantitative tools allows for quantitative analysis and comparisons. These comparisons can, for example, consist of different locations and their risk exposure, different business segments in the company, benchmarks between business segments of different companies or different technologies, methods or techniques. As an example, quantitative methods can be used to measure the impact of risk mitigation actions such as hedging price risks or diversification of the supplier structure. On that basis, the efficiency of alternative risk mitigation actions can be compared both from ex ante and ex post perspectives.

Thirdly, thanks to rapid advancements in computer science and technology, the use of quantitative tools has become easier. Data processing, combining, complex modelling and calculations can be performed within seconds and even less proficient users can perform sophisticated analyses using current software programs and tools.

Gap 3: Supply Chain Risk Management as a Dynamic Process Re-evaluation of the residual risk was not fully integrated in the supply chain risk management process in the automotive industry.

According to ISO 27001, the residual risk is the remaining risk once the initial risk is mitigated. In our interviews, we found that most firms practice the four consecutive risk management steps of identification, analysis, mitigation choice, mitigation execution. Most of our interviewed organisations do not integrate the re-evaluation of the residual risk after mitigation strategy execution into the dynamic risk management process. It is not always possible or realistic to completely eliminate all the risks. Therefore, some *residual* risks remain at a certain level.

In our research, we have seen that the automotive industry has implemented routine strategies for monitoring the development of their residual risks. This practice is more prevalent with OEMs compared to first- and second-tier suppliers. Smaller suppliers in particular do not show such recurring practices.

The key consideration is that the risk mitigating effort actually reduces the corporate impact and does not just move the risk to another similarly high level quadrant. Thus, once the mitigation step has been identified and executed, the risk should be reassessed, taking into account its effect and frequency. Its impact should then be recalculated to see whether the mitigation and execution effort has really made an acceptable difference. The mitigation efforts are often resource intensive and a major outlay for little or no residual risk should be challenged.

Basically, there are three options once the residual risks have been identified. First, if the risk is below an acceptable level, managers need to formally accept them. Second, if the risks are above an acceptable level, then the company needs to find new and better ways to mitigate them—this also means that reassessment of the residual risks is essential. Finally, if the risks are above the acceptable level, and if the cost of decreasing them would be higher than the impact itself, then the recommendation to managers would be that these risks should be accepted even if they are high. A systematic approach such as this ensures that managers are involved in the most important decision-making and that nothing is overlooked. After all, the management board is not only responsible for the bottom line of the company, but also for its viability (Kosutic 2012).

Practical Recommendations

The Three-Circle Model for Risk Identification and Assessment

Our first recommendation focuses on solving Gap 1. The risk identification and assessment stages are particularly critical to successful risk management as they detect organisational exposure to uncertainty (Neiger et al. 2009). Without a comprehensive knowledge of existing and potential disruption, appropriate actions for avoiding supply chain vulnerability will lose its evidential base. There are various approaches or models for tackling supply chain risks.

Based on prior research and practical approaches, we recommend that companies adopt the Three-Circle Model for supply chain risk identification for multi-tier supplier management, as illustrated in Fig. 3. The Three-Circle Model identifies and analyses risks from three angles, namely location, industry and the supply chain partner. It also has unique practical implications. Managers can use it to develop company-specific risk management programmes and to create strategies and management responses that can influence and enhance their relationships with multi-tier suppliers. The model is easy to implement and facilitates proactive supplier risk management.

In practice, the items for examination should best reflect the purpose of multi-tier supply chain risk management. The time gap between two updates can be large. As other parameters such as financial performance and logistics can be more prone to fluctuations, they will need updating more frequently.

Compared to the other risk models of supplier risk assessment, the Three-Circle Model has three unique advantages. Firstly, it provides an operational and holistically balanced approach to tackling risk types, drivers and their correlations in the context of multi-tier supplier management. Vast amounts of data and other corporate resources are needed to generate the risk map of multi-tier supplier demands.

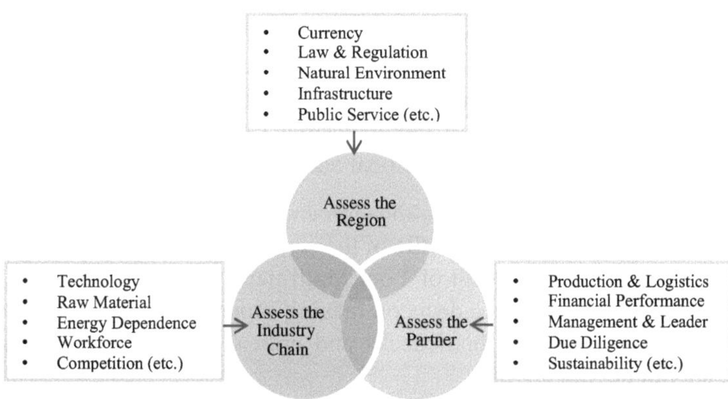

Fig. 3 Three-circle model: mapping the risk of your multi-tier supply chain partner

For automotive OEMs, the base of the pyramid tier system consists of tier four suppliers who form and perhaps own much of the intellectual capital that is required to produce excellent components. Many of those lower-tier suppliers own patents, have customised machinery and are family owned. Working with these lower-tier suppliers is the key to making the necessary changes in a timely and efficient manner. However, the current frameworks of supplier risk assessment do not fully integrate the lower-tier suppliers in a systematic way that takes into account both the regional and industry chain-related context.

Secondly, the three-dimensional view allows a more nuanced assessment of the major risk categories for each supplier and sub-supplier relationship. Although the Three-Circle Model is organised into overlapping segments, it can be operationalised based on a recurring and responsive process, which compensates for shortcomings in the identification process, such as the overlapping or bypassing of elements.

Lastly, the Three-Circle Model also pays additional attention to the assessment of due diligence. According to our interviews, less than 50 % of the interviewed companies have due diligence in place to alert them to supply chain risk. The supplier risk model is often built upon modelling historical financial performance and trends as an indicator of future supplier financial performance. However, supply chain operations change more quickly than financial projections. Especially in emerging countries, financial and accounting practices and standards vary, which makes a standardised, formulaic risk-modelling approach based purely on financial calculations less reliable.

The appropriate nature and extent of due diligence is dependent on individual circumstances and affected by factors such as the size of the organisation, the location of its activities, the situation in a particular country and the sector and nature of the products or services involved. These factors can be reflected by two dimensions of the Three-Circle Model, namely Region and Industry Chain. These challenges can be met in a variety of ways, including but not limited to (OECD 2013):

- Industry-wide cooperation in building capacity to conduct due diligence
- Cost sharing within the industry for specific due diligence tasks
- Participation in initiatives on responsible supply chain management
- Coordination between industry members who share suppliers
- Cooperation between upstream and downstream companies
- Building partnerships with international and civil society organisations

After the completion of risk identification and risk assessment, we recommend that managers delineate the risk profile of the supply chain partner in a visual way, as shown in Fig. 4—Risk Portrait of Your Multi-tier Supply Chain Partner.

The Risk Portrait serves as a decision-making support tool that enables companies to more easily identify bottlenecks and critical nodes within the risks for multiple suppliers.

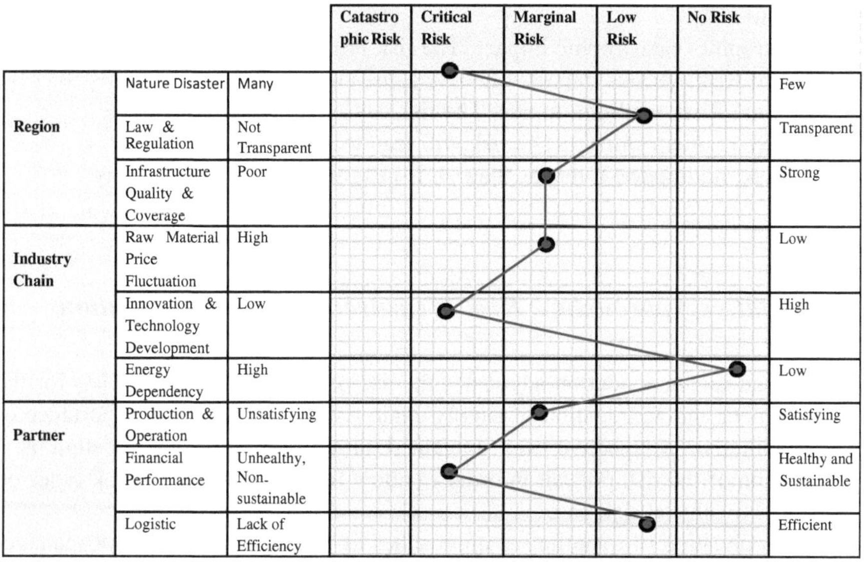

Fig. 4 Risk portrait of your multi-tier supply chain partner (Example)

To illustrate this, take one automotive OEM as an example. Suppose this OEM has only two second-tier suppliers which produce a critical component. The Risk Portraits show that the highest risk with the potential catastrophic impact faced by both of them is law and regulation. Once this critical information is acquired by the OEM, the OEM can decide to utilise its capabilities and resources to develop either proactive or reactive strategies to avoid or minimise the negative impact caused by risks of the second-tier suppliers.

In this example, we only take two suppliers as an example. However, if the number of suppliers is scaled up to a 100 or a 1000, it will become difficult for the OEM to compare the differences and similarities of the risk portraits manually. Thus, the next step in developing the conceptual Risk Portrait Model is to add software programs that allow assigning a value to the risk and the risk portrait automatically based on available indicators, so that multiple risk portraits can be easily compared and developments in one class can be observed and followed up on.

The parameters listed in Fig. 4 are for demonstration purposes. The choice of parameters below each category may differ from company to company. The result of the Risk Portrait needs to be updated regularly in order to draw comparisons between the historical and the current evaluations of a specific supply chain partner or between two or more partners. Moreover, this specific assessment allows the derivation of more nuanced and specific countermeasures for tackling risks compared to simply multiplying probability by business impact to derive a standard risk indicator.

Definition

- Catastrophic: catastrophic impact. The risk brings about discontinued business.
- Critical Risk: serious impact. The risk causes long-term difficulties.
- Marginal Risk: medium impact. The risk causes short-term difficulties.
- Low Risk: minor impact. The risk will lead to single small losses.
- No Risk: no impact. It means the risk is insignificant for the company

Quantitative Approaches: Risk Measurement and Aggregation

Our second recommendation advocates the use of quantitative approaches for the measurement and aggregation of supply chain risks. The emerging importance of the quantitative perspective was explained in Gap 2. The quantification is a progression of the risk portrait in order to allow for a comparison of risk types on a quantified value at risk basis.

The application of quantitative approaches as part of day-to-day risk management usually requires significant effort. It is therefore reasonable to prioritise supply chain risks according to their importance for the specific business and to consider the application of quantitative measurement only for those risks that receive top priority. In order to achieve this systematically, a Risk Map or Risk Portrait can be established, which reveals high-priority risks. Those risks are then represented by so-called risk factors, which can be directly observed as quantities (e.g. fluctuation of the EUR: USD exchange rate; change in the turnover volume of a certain product or the rise and fall of client defaults over a certain period of time).

Although intuitive, the practical application of risk mapping for supply chain risks should include safeguards. The pitfalls and avoidance strategies of risks are outlined in Table 3.

Quantitative Impact Assessment

As a result of the risk mapping procedure, standardised movements of risk factors representing highly prioritised risks can be assigned with their related (qualitative) impacts, as summarised in Table 3.

As the first step towards quantitative risk measurement, the above-mentioned impacts have to be quantified, or they need to be given a "price tag". For this purpose, the various contributions of losses associated with a certain risk factor movement (e.g. a supplier default or a shift in a particular market price) are determined taking into account all relevant parts of the supply chain. Some examples are illustrated in Fig. 5. As a quantitative reference for "loss", for instance, the future deviation of Earnings before Interest and Tax (EBIT) or the cash flow from a predetermined target figure can be used. The deviation will then refer to a certain period of time in the future, for example, a fiscal year.

Table 3 Risk mapping—pitfalls and avoidance strategies

"Unknown unknowns": failure to identify essential risks	"Known unknowns": misjudgement of the impact or probability of risks identified
Systematically analyse risks based on standardised risk categories for the business following a strict top-down approach (e.g. counterparty risks → client/supplier defaults)	Represent the risks identified in terms of a minimum set of main risk factors without pronounced correlations in order to limit the complexity of the analysis and to avoid implicit risk concentrations
Consider risks from a multi-tier supplier perspective, i.e. at least up to the second-tier supplier	Strictly separate the analysis of impact and probability
Collect information on risks from experts for each part of the supply chain (e.g. procurement, production, sales) based on a standard questionnaire	Assess (qualitative) impacts through standardised scenarios for the movement of all risk factors within a risk certain category (e.g. ±10 % shift in a certain market price or currency exchange rate)
Collect information from historical loss events for own business and for third parties exposed to comparable risks	Evaluate the (qualitative) probability of risk factor movements according to the standardised scenarios (e.g. low/medium/high) within a certain period of time
Permanently examine whether the risks which have led to historical loss events are properly represented in the risk map	

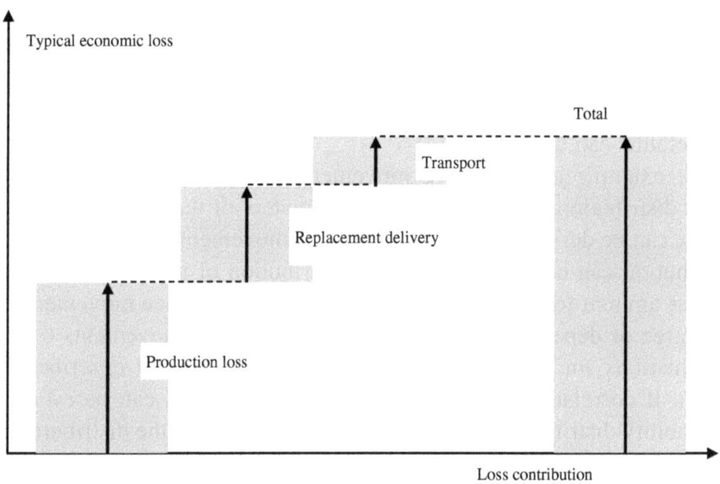

Fig. 5 Components of the typical economic loss related to a supplier default

For each loss contribution, a typical loss amount should be determined independently by at least two different experts. Where expert assessments vary significantly, expert interviews can be repeated if necessary.

Since the assessment of an impact remains inherently subjective in spite of the involvement of independent experts, it is important to verify the consistency of those assessments with historical loss events ("back-testing").

For simplicity, we assume a linear dependence between risk factor movements and the associated impact. This means that if an adverse market price movement doubles so too does the associated loss. Or if two clients default, the loss will be approximately twice as high as in the case of a single client default. This assumption has been made for simplicity or our illustration. Certainly the developments can also be concave or convex.

Modelling Joint Risk Factor Dynamics

As the final ingredient in the quantitative risk measurement, we have to find a model for probabilities (or more generally, probability distributions) of possible future risk factor movements. On that basis and taking into account the results of the previous section, the probability distribution for possible future loss events can be derived.

For example, the probability of a supplier default within 1 year might be estimated as $x\%$. The probability of default can be determined using statistical methods based on key information from the supplier, such as financial data (i.e. balance sheet figures), behavioural track records (delivery reliability, etc.) and information on its dependence on its own suppliers (e.g. second-tier suppliers).

Assuming that supplier defaults are statistically independent, the probability of more than one default, i.e. a specific number of defaults, is described by binomial distribution (Deutsch 2009). Given the typical loss amount associated with one supplier default, the probabilities of possible total losses resulting from several supplier defaults can be derived.

Another example is that future movements of prices are often described by lognormal distributions (Deutsch 2009). Parameters of the distributions, i.e. price volatilities, can be derived from historical price movements. In turn, the associated loss distribution can be derived from the distribution of market price movements and the loss amount for the firm associated with a given price movement.

The degree of dependence between future risk factor movements (e.g. market price fluctuations for two different commodities) is usually described by their correlation. If correlations between risk factor movements can be estimated, the joint probability distribution can be specified. For instance, the distribution for two different commodities defines the probability of any given simultaneous price movements for those commodities.

Quantitative Risk Measurement

Using ingredients described in the previous section, we can derive the probability distribution of losses caused by future movements of a single risk factor over a certain period of time, as demonstrated in Fig. 6.

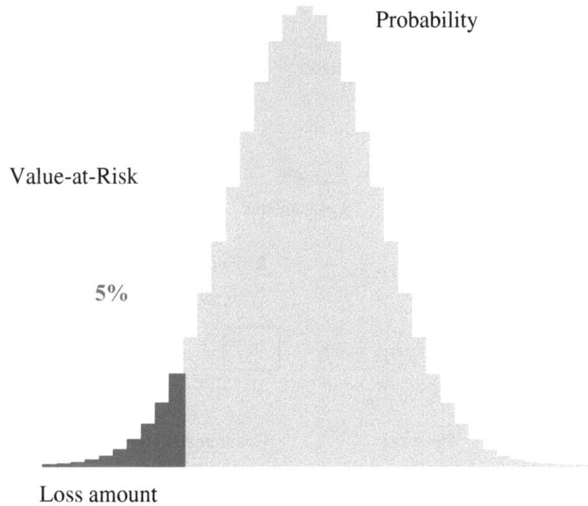

Fig. 6 Loss distribution and value-at-risk for a single risk factor

The risk involved in distribution can be described by a single risk measure. A common choice is Value-at-Risk (Deutsch 2009) which denotes, with a certain confidence, the loss level which will not be exceeded (e.g. 95 %) within a certain period of time. In other words, the actual loss will exceed the Value-at-Risk with a probability of 5 %, i.e. in one out of 20 observations.

Value-at-Risk can be estimated analytically using parametric probability distributions. If there is no explicit analytical representation, then Monte Carlo techniques can be used, for instance, to simulate possible outcomes for future loss based on random numbers (Finke et al. 2010). For example, the Value-at-Risk for 95 % confidence can be estimated as the fifth percentile of the simulated outcomes for the future loss.

Aggregation of Supply Chain Risks

Aggregated loss distributions for multiple risk factors can be derived from the related distributions for single risk factors together with the correlations of these factors. Due to the complex structure of multi-factor distributions, it is uncommon to model those distributions directly based on parametric functions. Instead, aggregated risk measures such as Value-at-Risk are derived using Monte Carlo simulation techniques.

The aggregation of risks can be executed along different dimensions, in particular through type of risk (e.g. market price risk, counterparty risk, etc.) or by business unit, as illustrated in Fig. 7.

Thus, it is possible to analyse and track down contributions to supply chain risk from all areas. The knowledge gained in this process can then be used as a key ingredient for strategic decision-making—from the choice of efficient risk mitigation techniques such as hedging to business and investment planning.

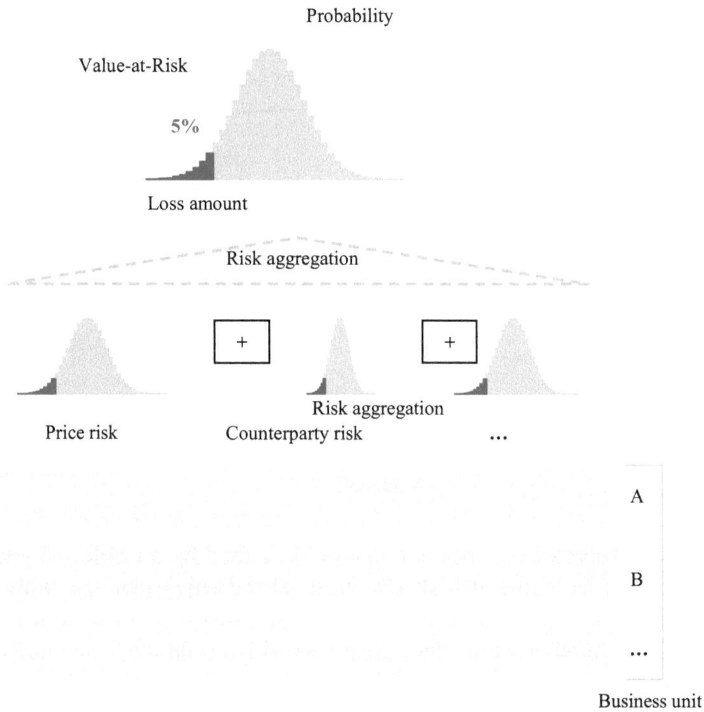

Fig. 7 Aggregation of supply chain risks

Risk Management Process: Multi-tier Supply Chain Risk Management Circle

Our third recommendation focuses on solving Gap 3. The model recommended here concentrates on integrating the re-evaluation of the residual risk in the supply chain risk management. This model can be applied in both of the industries studied.

We have developed a five-stage framework for multi-tier supplier risk management, incorporating risk identification, risk assessment, risk mitigation strategy development, risk management execution and residual risk re-evaluation. It builds on Ritchie and Bridley's conceptual approach and the approach of the Association of Insurance and Risk Managers (AIRMIC). We argue that the supply chain risk should be managed as a dynamic process. A dynamic process requires that risk management objectives, strategies, tactics and priorities fit into the dynamic external and internal business environments. This closed feedback loop provides three suggestions for practice.

To implement this model, companies in both industries researched here, especially those with multiple product lines, divisions or brands, need to firstly define their supply chain risk management so that they are distinct from a one-for-all risk management strategy and implementation plan. To a large extent, the majority of

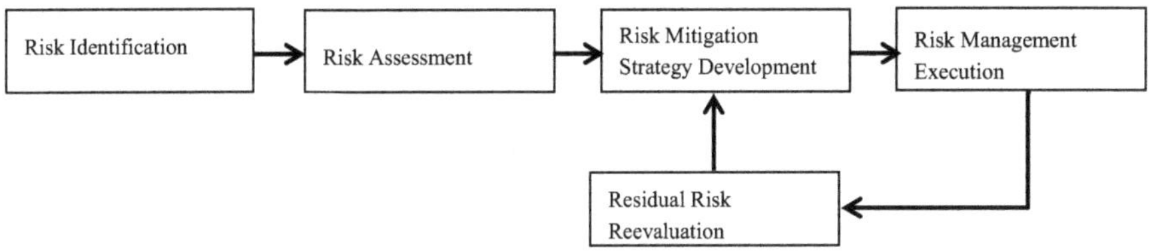

Fig. 8 Multi-tier supply chain risk management circle

the companies in our case study belong to this category, as they have multiple product lines and an internal customer base.

Secondly, to implement the tool, managers should know that the objective of supply chain management may evolve with time. Due to organisational and environmental changes, companies, even with rather simple product lines, will be exposed to different supply chain risks at different times. Thus, risk management needs to be regarded as an organic process. For the majority of our interviewed companies, cross-border geographic expansion and executive restructuring has taken place in the last five years. For instance, as one of the world's leading suppliers of high-precision metal components for the automotive industry, one company interviewed opened its wholly owned subsidiary in Suzhou in 2009. As a result, the company has had the opportunity to supply numerous European customers at one of its major foreign sites and also has access to new customers in Asia. Therefore, it is important to update supply chain risk management strategy and practice in order to reflect the dynamic internal and external operating environment.

Thirdly, risk management procedure should be a closed-loop circle, as shown in Fig. 8. In this dynamic process, supply chain risk management starts at risk identification but does not end with the execution of risk management strategies and tactics. In our model, we advocate risk management execution being the starting point of the next round of risk re-evaluation.

Conclusion

In this empirical research, we have studied supply chain risk management in the automotive industry. Our interview findings include, firstly, that at present, industrial environment of supplier risk management is often simplified as first-tier supplier risk management; secondly, quantitative supply chain risk management approaches are not widely applied in practice; and thirdly, practice re-evaluation of the residual risk is not commonly practiced in the supply chain risk management process. In addition, the practice of due diligence is not fully integrated into supplier identification and evaluation procedures. Furthermore, integrative risks, meaning the spread and amplification of risks upstream and downstream of the supply chain, have not yet been managed to the highest degree possible.

As an important part of supply chain management systems, we advocate that a holistic and integrative approach to multi-tier supply chain risk management is no longer just an option, but is recognised as being decisive as globalisation, outsourcing and offshoring stretch the supply chain to its limit. The significant advantage of multi-tier visibility and collaboration is that it enables manufacturers to synchronise planning and coordinate execution activities based on information from all relevant stakeholders along the chain as a way of efficiently managing supply chain risks or disruptions.

The strategies and approaches for companies to avoid or reduce multi-tier supply chain risks vary. There are various factors which contribute to these differences. In spite of the complexity of multi-tier supply chain risk management, there are a growing number of successful examples that demonstrate its feasibility in practice. Current technologies such as radiofrequency identification, enterprise resource planning and general packet radio service will become important tools for supply chain network management (Tang 2006; Vanany et al. 2009), and future technologies are expected to make a big impact on visibility and traceability of supply chain networks (Ghadge et al. 2012).

Based on both empirical and scientific research, we have introduced three conceptual models to help companies to avoid supply chain risks and to seize potential opportunities. Firstly, the Three-Circle Model helps managers to identify the risks of a multi-tier supply chain partner. Secondly, the Risk Portrait of the Multi-tier Supply Chain increases the risk visibility and comparability of the selected multi-tier supplier. Thirdly, the quantitative approach to assessing multi-tier supply chain risks includes five basic steps, namely prioritisation of supply chain risk by using tools such as Risk Mapping, quantification of the risk impact, modelling joint risk factor dynamics, calculating risk probability distribution and aggregation of supply chain risks.

Last but not least, there exists no simple solution for managing multi-tier supply chain risks. And there is no simple way to imitate the best supply chain. How supply chain operation and risk should be managed relies on internal corporate factors such as corporate strategy, core competitiveness and corporate resources, as well as external environment factors such as competition and regulation. The evidence seems compelling: only an appropriate supply chain management scheme can enable a company to reach a balanced set of metrics that are aligned with the ultimate corporate goal and the firms' risk appetite.

Acknowledgements The authors are grateful to Dr. Tilman Huhne for his valuable suggestions for improving the earlier version of the paper. We also appreciate Vicky English's editorial support.

Bibliography

Alvarenga, C. (2012). *Why quants should manage your supply chain risk*. Harvard Business School Publishing. Accessed July 23, 2015, from http://blogs.hbr.org/2012/11/why-quants-should-manage-your/

Becks, R. (2010). The next big thing: Multi-tier supply chain management. Accessed July 23, 2015, from http://www.scdigest.com/experts/E2Open_11-08-11.php?cid=4840&ctype=content

Carey, W. P. (2005). Deep supplier relationships drive automakers' success. Accessed July 23, 2015, from http://knowwpcarey.com/article.cfm?aid=940

Childerhouse, P., & Towill, D. R. (2003). Simplified material flow holds the key to supply chain integration. *Omega, 31*(1), 17–27.

Choi, T. Y., Dooley, K., & Rungtusanatham, M. (2001). Supply networks and complex adaptive systems: Control versus emergence. *Journal of Operations Management, 19*(3), 351–366.

Christopher, M., & Lee, H. (2004). Mitigating supply chain risk through improved confidence. *International Journal of Physical Distribution and Logistics Management, 34*(5), 388–396.

Deutsch, H. P. (2009). *Derivatives and internal models* (4th ed., pp. 247–331). London: Palgrave Macmillan.

Eisenhardt, K. M. (1989). Building theories from case study research. *The Academy of Management Review, 14*(4), 532–550.

Financial Times. (2013). Can companies know every supplier? Accessed July 23, 2015, from http://www.ft.com/intl/cms/s/0/f1e99190-7a03-11e2-9dad-00144feabdc0.html#axzz2hmahz8wK

Finke, G. R., Sprödt, A., & Plehn, J. F. (2010). A comparison of quantitative methods in supply chain risk management—Benefits and drawbacks. Accessed July 23, 2015, from http://www.pomsmeetings.org/ConfPapers/015/015-0668.pdf

Frohlich, M. T., & Westbrook, R. (2001). Arcs of integration: An international study of supply chain strategies. *Journal of Operations Management, 19*(2), 185–200.

Gartner Inc. (2011). Predicts 2012: Supply chain predictions: Talent, risk and analytics dominate. Accessed July 23, 2015, from http://www.gartner.com/id=1853218

Ghadge, A., Dani, S., & Kalawsky, R. (2012). Supply chain risk management: Present and future scope. *The International Journal of Logistics Management, 23*(3), 313–339.

Harland, C., Brenchley, R., & Walker, H. (2003). Risk in supply networks. *Journal of Purchasing and Supply Management, 9*(2), 51–62.

Hendricks, K. B., & Singhal, V. R. (2003). The effect of supply chain glitches on shareholder wealth. *Journal of Operations Management, 21*(5), 501–522.

Jüttner, U. (2005). Supply chain risk management: Understanding the business requirements from a practitioner perspective. *The International Journal of Logistics Management, 16*(1), 120–141.

Kajüter, P. (2003). Instrumente zum Risikomanagement in der Supply Chain. In W. Stölzle & A. Otto (Eds.). *Supply Chain Controlling in Theorie und Praxis: Aktuelle Konzepte und Unternehmensbeispiele* (pp. 107–135). Wiesbaden.

Kern, D., Moser, R., Hartmann, E., & Moder, M. (2012). Supply risk management: Model development and empirical analysis. *International Journal of Physical Distribution and Logistics Management, 42*(1), 60–82.

Knemeyer, A. (2009). Proactive planning for catastrophic events in supply chains. *Journal of Operations Management, 27*(2), 141–153.

Kosutic, D. (2012). Risk assessment and risk treatment methodology. Accessed July 23, 2015, from http://www.iso27001standard.com/en/documentation/Risk-Assessment-and-Risk-Treatment-Methodology

Levitt, T. (1983, May–June). The globalization of markets. *Harvard Business Review, 61*, 92–102.

Neiger, D., Rotaru, K., & Churilov, L. (2009). Supply chain risk identification with value-focused process engineering. *Journal of Operations Management, 27*(2), 154–168.

Norrman, A., & Jansson, U. (2004). Ericsson's proactive supply chain risk management approach after a serious sub-supplier accident. *International Journal of Physical Distribution and Logistics Management, 34*(5), 434–456.

OECD. (2013). OECD due diligence guidance for responsible supply chains of minerals from conflict-affected and high-risk areas (2nd edn.). OECD Publishing. Accessed July 23, 2015, from doi:10.1787/9789264185050-en

Paulsson, U. (2007). On managing disruption risks in the supply chain—The DRISC model. Accessed July 23, 2015, from http://www.husdal.com/2009/03/02/managing-disruption-risks-in-the-supply-chain-the-drisc-model/

Peck, H. (2005). Drivers of supply chain vulnerability: An integrated framework. *International Journal of Physical Distribution and Logistics Management, 35*(4), 210–232.

Porter, M. E. (1985). *The competitive advantage: Creating and sustaining superior performance.* New York: Free Press.

Ritchie, B., & Brindley, C. (2007). Supply chain risk management and performance: A guiding framework for future development. *International Journal of Operations and Production Management, 27*(3), 303–322.

Romeike, F., & Hager, P. (2009). *Erfolgsfaktor Risiko-Management 2.0* (2nd ed.). Wiesbaden: Gabler.

Sodhi, M. S., Son, B., & Tang, C. S. (2012). Researchers' perspectives on supply chain risk management. *Production and Operations Management, 21*(1), 1–13.

Strauss, A., & Corbin, J. (1989). *Tracing lines of conditional influence: Matrix and paths.* Paper Delivered at the Annual Meetings of the American Sociological Society, San Francisco, CA.

Svensson, G. (2004). Key areas, causes and contingency planning of corporate vulnerability in supply chains: A qualitative approach. *International Journal of Physical Distribution and Logistics Management, 34*(9), 728–748.

Tang, C. S. (2006). Robust strategies for mitigating supply chain disruptions. *International Journal of Logistics Research and Applications, 9*(1), 33–45.

Tang, C. S., & Zimmerman, J. D. (2009). Managing new product development and supply chain risks: The Boeing 787 case. *Supply Chain Forum, 10*(2), 74–86.

Taylor, D. H. (1999). Measurement and analysis of demand amplification across the supply chain. *International Journal of Logistics Management, 10*(2), 55–70.

The New York Times. (2013). Japan's role in making batteries for Boeing. Accessed 23 July 2015, from http://www.nytimes.com/2013/01/26/business/selection-of-the-boeing-787s-battery-maker-raises-questions.html?pagewanted=all&_r=0

Thun, J. H., Marble, R. P., & Silveria-Camargos, V. (2007). A conceptual framework and empirical results of the risk and potential of just in sequence—A study of the Germany automotive industry. *Journal of Operations and Logistics, 1*(2), I.1–I.13.

Trkman, P., & McCormack, K. (2009). Supply chain risk in turbulent environments—A conceptual model for managing supply chain network risk. *International Journal of Production Economics, 119*(2), 247–258.

Vanany, I., Zailani, S., & Pujawan, N. (2009). Supply chain risk management: Literature review and future research. *International Journal of Information Systems and Supply Chain Management, 2*(1), 16–33.

Waters, D. (2007). *Supply chain risk management: Vulnerability and resilience in logistics.* London: Kogan Page.

Winter, K. (2012). Predicting 2012: Talent trends in supply chain process model. Accessed July 23, 2015, from http://www.logisticsexecutive.com/news_and_insights/media_and_articles/2012/predicting_2012_-_talent_trends_in_supply_chain

Temporal Shifts in Freight Traffic: Estimating the Potential to Improve Traffic Quality on German Highways with Temporal Shifts of Heavy Traffic

Karin Menges and Manfred Boltze

Introduction

The road transport performance in Germany has increased significantly since the 1990s. Most likely, this trend will continue in the following years (SRU 2012). An increase of freight transport volume by one-third is estimated for roads in the Federal State of Hessen until 2030 (IVM 2010). However, traffic infrastructure in Hessen is already operating up to capacity during peak periods due to high traffic volumes, which causes congestion and high levels of pollution (HMWVL 2013). It can be assumed that the impact of capacity limitations will become even more obvious in future since necessary investments in transport infrastructure will most probably not be realised due to a shortage of public funds (Ahrens and Kabitzke 2011). Investments of 7.2 billion euros p.a. are required in order to maintain the German transport infrastructure (Daehre-Komission 2012). The underinvestment may lead to an increasing number of motorway sections operating under insufficient quality levels (ADAC 2012). Therefore, the existing transport infrastructure has to be used in a more efficient way.

Due to long truck convoys on motorways or traffic disruptions caused by overtaking trucks, freight transport is often considered as a major cause of capacity constraints in road traffic. Thus, it is necessary to seek for solutions to influence the freight transport demand. A promising approach is the concept of a freight transport

K. Menges (✉) • M. Boltze
Chair of Transport Planning and Traffic Engineering, Technische Universität Darmstadt, Otto-Berndt-Straße 2, 64287 Darmstadt, Germany
e-mail: menges@verkehr.tu-darmstadt.de; boltze@verkehr.tu-darmstadt.de

E. Abele et al. (eds.), *Dynamic and Seamless Integration of Production, Logistics and Traffic*, DOI 10.1007/978-3-319-41097-5_8

demand management within traffic management (see Rühl and Boltze 2016). Influencing time and modal choice are established approaches within the concept of mobility management for passenger transport. Accordingly, the optimisation of the impacts of freight transport should be achieved by influencing freight transport demand. In addition to measures for traffic reduction, spatial and modal traffic shift or traffic routing, the temporal shift of freight traffic is a considerable option in order to relieve the existing infrastructure and to reduce peak loads during rush hour. However, the question arises immediately whether and to what extent there are potentials for a temporal shift of freight traffic on motorways. Besides, it is necessary to determine the impacts of corresponding freight transport demand management measures.

In the following sections, the theoretical potential of a daytime-dependent shift of freight traffic on federal motorways will be analysed. Therefore, the following aspects will be investigated:

- To what extent do overall traffic peak hours and heavy goods vehicle (HGV) traffic peak hours (according to FGSV 2012 defined as the 60-min time period with the highest traffic volume within a day) overlap during the course of the day? To what extent do time periods with high traffic volumes (defined here as traffic volumes higher than 90 % of the peak volume) overlap?
- Is it theoretically possible to achieve a sufficient quality of traffic flow (Level of Service—LOS D) during periods of high traffic volumes by temporally shifting HGV traffic only?
- Is it possible to cluster HGV traffic patterns of different weekdays, so that generally applicable statements can be made with regard to specific weekdays?
- Are effects of existing measures for the temporal shift of HGV traffic identifiable by analysing the truck ban during vacation time which is in force on Saturdays from July to August on specific motorways and federal highways in Germany?

Based on the results, conclusions should be drawn on the future role of measures for temporal shifts of HGV traffic within the framework of traffic management and specifically within demand management.

Existing Measures to Influence the Time Choice

In the field of passenger transport, influencing the time choice of users has already been successfully implemented. Especially in public transport, there are distinct daily fluctuations in demand. A high number of trips with educational or professional purposes can be noted during peak hours and especially in the morning. Usually, this leads to very high load factors or even an overloading of public transport vehicles.

During non-peak hours, however, large parts of the capacity are not utilised. For a long time, the adjustment of the supply side has been the common reaction towards a rise at the demand side. Since such measures will most probably not be considered as priority in the future due to the funding problem, adjusting the demand side becomes more and more important. In particular, influencing time choice is regarded as a possible solution. Relevant measures are, for example, the time shift or staggering of teaching times at schools and commutation tickets with reduced prices and the restriction of an exclusive validity during non-peak hours (Kittler and Boltze 2011).

As mentioned above, with regard to motorised private traffic, an overloading of transport systems can be noticed regularly during peak hours. For passenger transport, mobility management offers a range of measures aiming to shift traffic temporally (e.g. city tolls, parking fees, etc.). Also for freight transport, measures for influencing the time choice are used in order to influence traffic flow in a positive way. A popular way of influencing time choice of inner-city delivery traffic operators is the implementation of delivery time windows. Additionally, measures such as inner-city bans for through-traffic or nighttime bans for certain vehicles have been implemented because of rising noise and air pollution. These measures also influence at least partially the time choice of freight transport. In foreign countries (e.g. in the cities of Stockholm, Singapore), staggered city toll concepts are applied to influence the time choice.

Furthermore, dynamic toll concepts are going to be used on motorways abroad (e.g. in the United Kingdom and the Netherlands) in order to influence the time choice of freight transport. In Germany, there are restrictions for HGV traffic such as the truck ban on Saturdays during the major vacation time (between July and August) or the Sunday truck ban. These bans aim at harmonizing the traffic flow. The truck ban during vacation time is valid at selected motorway sections and highways and is to be obeyed by trucks with a permissible total weight of more than 7.5 t as well as trailers behind trucks on Saturdays from July 1 to August 31 between 7 a.m. and 8 p.m. Combined transport operations (Rail/Road and Port/Road) are excluded from the ban, as well as transport operations with the purpose of transporting perishable food (fresh milk, fresh meat, related products and fresh fruits). Besides, empty runs connected to the transport of perishable food are also excluded (German Road Traffic Regulations: Strassenverkehrs-Ordnung StVO § 30, 3 and 4).

By comparing measures influencing the time choice in public transport with those in freight transport, it becomes obvious that the measures used in public transport relate to a shift of demand peaks within a period of few hours, whereas in freight transport, with the exception of delivery time slots, only measures with a validity of several hours (e.g. driving bans lasting all day long) have usually been implemented regarding influencing time choice. However, in principle, implementing measures in freight transport to influence the volume in short term would also be desirable in order to smoothen overall traffic demand during peak hours.

Methodology

Study Area and Counting Sites

Within this study, only truck traffic loads (as a major part of freight transport) will be considered in order to determine the theoretical potential for a temporal shift of freight transport. The analysis and study on effects of the truck ban during vacation time, as an example for an existing measure to influence time choice of freight transport, is based on traffic data from six automated counting sites operated by the Bundesanstalt für Straßen- und Verkehrswesen (BASt) in the central German region Frankfurt RheinMain.

The chosen counting sites represent sections with high traffic loads. The analysis of the average weekly patterns of the year 2011 (without vacation time) showed either LOS E in both or LOS F in one driving direction. Besides, two of the selected counting sites are motorway sections affected by the truck ban during vacation time. By comparing these selected counting sites with four counting sites being not affected by the truck ban during vacation time, it is possible to estimate the effectiveness of existing measures.

Processing of Counting Site Data for Further Analysis

Raw data generated by counting sites provide information about the hourly traffic volumes on various road sections differentiated into nine vehicle types in accordance with BASt (2013) (see Table 1).

The following vehicle types were grouped into common categories in order to develop traffic volume patterns and to analyse the potential for temporal shift:

- Category 1: Passenger Cars
  (Passenger Cars, Passenger Cars with trailers, motorcycles, not classifiable motor vehicles)
- Category 2: Buses
- Category 3: Vans and
- Category 4: Trucks
  (HGV > 3.5 t without trailer, HGV > 3.5 t with trailer and tractor trailers)

Table 1 Differentiation of vehicle types (BASt 2013)

Passenger car		Van	
Passenger car with trailer		HGV > 3.5 t without trailer	
Motorcycle		HGV > 3.5 t with trailer	
Bus		Tractor-trailer	
Not classifiable motor vehicle			

Conversion of Traffic Volumes into Passenger Car Units (PCU)

A conversion of traffic volumes on each direction into PCU was conducted to identify the influence of HGV traffic on traffic quantity and traffic quality. Since the cross sections addressed are located outside of intersections, an approach based on the instructions for execution of "Leitfaden zum Arbeitsstellenmanagement auf Bundesautobahnen" (see BASt 2011) was used. Hence, the function to determine the traffic volume q_{PE} (in PCU) on motorways expressed in Eq. (1) may be described as follows:

$$q_{PE} = q \cdot \left(1 + \frac{b_{SV}}{100} \cdot (f_G - 1)\right) \tag{1}$$

q_{PE} Traffic volume (PCU/h)
b_{SV} Proportion of heavy traffic (%)
(proportion of categories 2 and 4 of overall traffic)
f_G Terrain factor
(plane $< 2\,\% = 1.5$; hilly 2–4 % = 2; mountainous $> 4\,\% = 2.5$)

The proportions of heavy traffic for the calculation of the traffic volume (PCU/h) were determined as hourly data based on BASt and were implemented into the calculations at the corresponding position. Assuming slopes smaller than 2 % at the selected sites, a terrain factor $f_G = 1.5$ was used to calculate traffic volumes q_{PE} according to BASt 2011). Depending on the proportion of heavy traffic, an hourly varying extrapolation factor for trucks could be generated from Eq. (1). That factor was 1.5 on average.

Determination of Levels of Service (LOS) with Respect to Traffic Flow

Figure 1 shows not only the course of traffic volumes but also the different LOS concerning traffic flow. Therefore, the potential to improve the LOS by temporal shifts of HGV traffic can be demonstrated. The capacity of a single carriageway was estimated according to "Handbuch für die Bemessung von Straßenverkehrsanlagen" (HBS) (FGSV 2009) in order to determine present LOS. Since the selected sections are two-lane or three-lane carriageways within agglomerations, a maximum capacity of 4000 or 5700 veh/h is used. By making use of critical values of the load factor a ($A \leq 0.3, B \leq 0.55, C \leq 0.75, D \leq 0.9$ and $E \leq 1.0$)

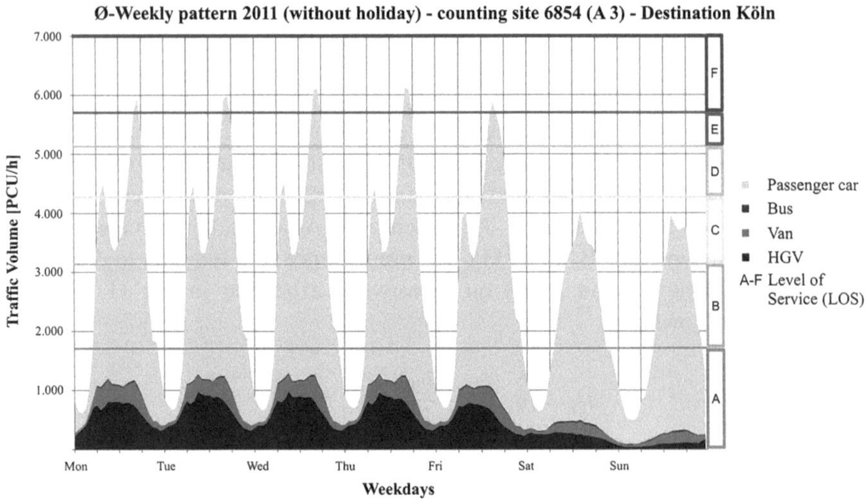

Fig. 1 Exemplary presentation of traffic volumes and levels of service of a specific counting site

and the capacities calculated above, the permissible traffic volumes (q_{zul}) for different levels of service can be determined.

The effect of gridlocks on the analysed traffic volume data, which could be shown simultaneously in speed data, is not considered because of simplification matters. Such effects would be relevant for studies of short time intervals within peak hours. However, for the purpose of qualitative analyses of daily traffic volume patterns, they can be neglected.

Generated Patterns

The data of counting sites were examined for both driving directions between 2006 and 2011 in order to analyse the theoretical potential for a daily temporal shift of HGV traffic and to analyse the effects of the existing truck ban during vacation time. The following types of patterns were generated from the raw data:

- Average daily patterns (Monday to Sunday): These patterns serve for the analysis of the overlapping of overall traffic peak hours and HGV traffic peak hours. In addition, they are used for the analysis of the overlapping of periods of high traffic volumes in general. Furthermore, the patterns provide information on the levels of service and the traffic volume that needs to be shifted to achieve at least a sufficient Level of Service (LOS D).
- Average daily patterns of HGV traffic (Monday to Sunday): Analysing these patterns can help to determine whether HGV traffic volume patterns of different weekdays can be clustered.

- Average patterns of Saturdays affected and not affected by the truck ban during vacation time: These patterns serve for the purpose of analysing the effects of the existing measures for temporal shift of HGV traffic.

Results

Overlapping of Peak Hours and Overlapping of Periods with High Traffic Volumes

The average daily patterns of all counting sites were analysed to check whether peak hours of overall traffic and HGV traffic occur simultaneously, in order to identify the potential overlapping of peak hours of overall traffic and HGV traffic. Additionally, it was examined if there is an overlapping of periods with high traffic volumes. High traffic volumes are defined as traffic volumes above a minimum level of 10 % less than the absolute peak volume (grey-hatched area) (see Fig. 2 for exemplary daily patterns).

Basically, the analysis has shown that peaks in overall traffic are significantly more intense with much higher traffic volumes (due to traffic volumes originating from passenger cars) than peaks in HGV traffic. The example in Fig. 2 demonstrates that periods of high volumes of overall traffic last about 1–3 h. Some counting sites show both peaks in the morning and in the afternoon, whereas other counting sites exhibit peak volumes only in the morning or in the afternoon. However, for HGV

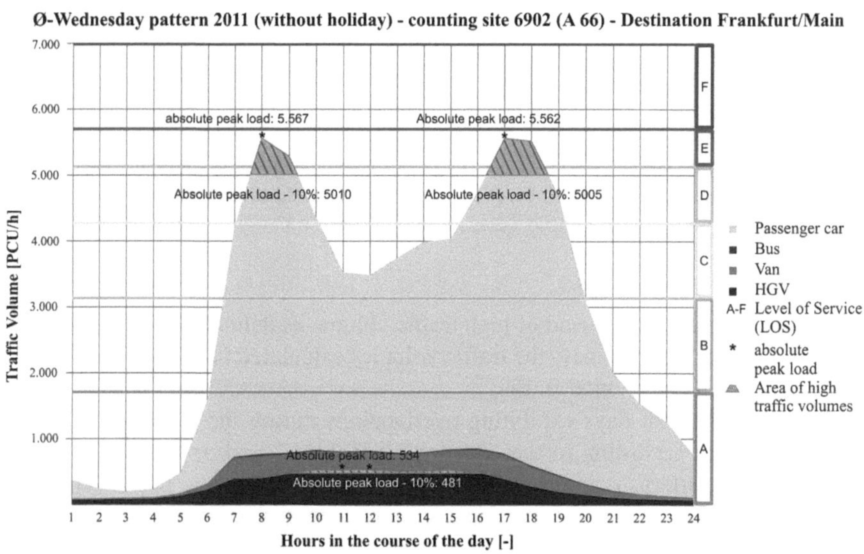

Fig. 2 Determination of the overlapping of peak hours and hours of high traffic volumes

Table 2 Proportion of overlappings of absolute peak hours during 2006–2011 (x_{WD}; x_{WE})

Monday–Friday (Working days: WD)		Saturday–Sunday (Weekend: WE)	
$x_{WD} = \frac{\sum_{i=1}^{6}(\overline{a}_i+\overline{b}_i+\overline{c}_i+\overline{d}_i+\overline{e}_i)}{30} \times 100\,\%$		$x_{WE} = \frac{\sum_{i=1}^{6}(\overline{f}_i+\overline{g}_i)}{12} \times 100\,\%$	
i	Years	i	Years
$\overline{a}$ to $\overline{e}$	1 in case of an overlapping of absolute peaks on average $\overline{a}$ = Mondays, $\overline{b}$ = Tuesdays, $\overline{c}$ = Wednesdays, $\overline{d}$ = Thursdays, $\overline{e}$ = Fridays; otherwise 0	$\overline{f}$ to $\overline{g}$	1 in case of an overlapping of absolute peaks on average $\overline{f}$ = Saturdays, $\overline{g}$ = Sundays; otherwise 0
30	Number of average working days investigated (5 Weekdays × 6 Years)	12	Number of average Saturdays and Sundays investigated (2 Days during weekend × 6 Years)

Table 3 Proportion of overlappings of periods of high traffic volumes (peak hour minus 10 %) during the investigation period from 2006 to 2011 (y_{WD}; y_{WE})

Monday–Friday (Working days: WD)		Saturday–Sunday (Weekend: WE)	
$y_{WD} = \frac{\sum_{i=1}^{6}(\overline{h}_i+\overline{i}_i+\overline{j}_i+\overline{k}_i+\overline{l}_i)}{30} \times 100\,\%$		$y_{WE} = \frac{\sum_{i=1}^{6}(\overline{m}_i+\overline{n}_i)}{12} \times 100\,\%$	
i	Years	i	Years
$\overline{h}$ to $\overline{l}$	1 in case of an overlapping of periods with high traffic volumes on average $\overline{h}$ = Mondays, $\overline{i}$ = Tuesdays, $\overline{j}$ = Wednesdays, $\overline{k}$ = Thursdays, $\overline{l}$ = Fridays; otherwise 0	$\overline{m}$ to $\overline{n}$	1 in case of an overlapping of periods with high traffic volumes on average $\overline{m}$ = Saturdays, $\overline{n}$ = Sundays; otherwise 0
30	Number of average working days investigated (5 Weekdays × 6 Years)	12	Number of average Saturdays and Sundays investigated (2 Days during weekend × 6 Years)

traffic there is only one period of high traffic volume as defined above. This period lasts for up to 8 h. Obviously, the traffic volumes calculated in PCU are a lot lower in HGV traffic than in other traffic.

The proportion of days exhibiting overlappings during the investigated period was determined according to the equations in Tables 2 and 3. The proportion of average daily patterns demonstrating an overlapping of absolute peaks (x) among the 504 examined (dependent on counting sites) and the proportion of patterns with overlapping periods of high traffic volumes (y) were deducted. Due to similar

Table 4 Analysis of the overlapping of peak hours and periods of high traffic volumes of overall traffic and HGV traffic

Counting site	Direction of travel	Weekdays	LOS of peak hours for 2006–2011	Overlapping of peak hours (x_{WD}; x_{WE}) (%)	Overlapping of periods of high traffic volumes (peak hour minus 10 %) (y_{WD}; y_{WE}) (%)
6853	Köln	Mo–Fr	D–E	0.0	100.0
		Sa–Su	C	0.0	33.3
	Würzburg	Mo–Fr	F	70.0	83.3
		Sa–Su	C	0.0	0.0
6854	Köln	Mo–Fr	F	0.0	19.0
		Sa–Su	C	0.0	25.0
	Frankfurt/ Main	Mo–Fr	F	10.0	90.0
		Sa–Su	C	0.0	0.0
6890	Frankfurt/ Main	Mo–Fr	D–E	3.3	60.0
		Sa–Su	B	33.3	58.0
	Eschborn	Mo–Fr	D–E	0.0	0.0
		Sa–Su	B	0.0	50.0
6902	Frankfurt/ Main	Mo–Fr	E	0.0	40.0
		Sa–Su	B	0.0	8.3
	Wiesbaden	Mo–Fr	E–F	0.0	96.7
		Sa–Su	C	0.0	8.3
6932	AD Mönchhof	Mo–Fr	E	0.0	43.3
		Sa–Su	C	0.0	8.3
	Darmstadt	Mo–Fr	E–F	0.0	50.0
		Sa–Su	C	0.0	8.3
6991	Rüsselsheim	Mo–Fr	E	0.0	50.0
		Sa–Su	B	0.0	8.3
	Mainz	Mo–Fr	E	0.0	86.7
		Sa–Su	B	0.0	0.0
	Mean values	Mo–Fr	–	6.9	59.9
		Sa–Su	–	2.8	17.3

traffic volumes, the working days from Monday to Friday (x_{WD}; y_{WD}) and the weekend from Saturday to Sunday (x_{WE}; y_{WE}) were clustered.

Table 4 shows the results of the analysis of a temporal overlapping of peak hours and periods of high traffic volumes of overall traffic and HGV traffic.

It can be summarised that only 7 % of the investigated daily patterns show an overlapping of peak hours of overall traffic and HGV traffic on working days. Overlappings of periods of high traffic volumes are recognisable in about 60 % of all patterns on working days. It must be mentioned that the period of overlapping is rather short in almost all cases and only last for about 1 h. Furthermore, the proportion of truck traffic on overall traffic is always very low with a maximum of 12 % (with regard to PCU) during the few hours of overlapping. This analysis of daily patterns clearly shows that passenger car traffic contributes to a large extent to the peak volumes observed. Especially, this is important in the case of the implementation of vehicle-dependent measures to improve traffic flow by influencing time choice according to the causative principle (dependent on the proportional amount on overall traffic volume).

It is not necessary to analyse the overlapping of peak hours and periods of high traffic volumes on Saturdays and Sundays in detail, since the generally low traffic volumes at the weekend reach average LOS B and C.

Temporal Shift of Traffic Volume Necessary to Achieve Level of Service D

The examination whether LOS D could be achieved only by a temporal shift of HGV traffic is deducted by using average daily patterns of all counting sites and years available. The required traffic volume (in PCU) to achieve a sufficient LOS is identified by analysing these patterns. Subsequently, the necessary percentage of HGV traffic to be shifted is calculated.

The hatched areas in Fig. 3 display the proportion of traffic volume exceeding LOS D for counting site 6854. In addition, the traffic volume which needs to be shifted and the traffic volume caused by HGV during this period are compared graphically in order to illustrate the proportion of HGV that needs to be shifted.

On 290 of the 504 days investigated (58 %) LOS D was exceeded at least once. The exceedance was of different duration and intensity. For further analysis, the

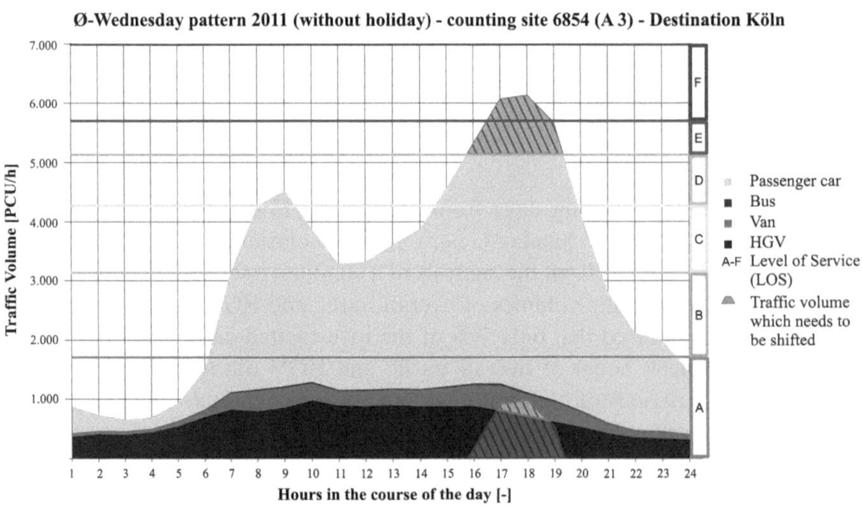

Fig. 3 Determination of the traffic volume to be shifted to achieve level of service D

Table 5 Evaluation of the shift of truck traffic to achieve level of service D

Counting site	Direction of travel	Number of hours exceeding LOS D among all ø-weekly patterns from 2006 to 2011	Number of hours with a necessary shift of more than 100 % of HGV traffic	Number of hours with a necessary shift of more than 75 % of HGV traffic
6853	Köln	4	0	0
	Würzburg	60	33	42
6854	Köln	95	44	61
	Frankfurt/ Main	97	62	63
6890	Frankfurt/ Main	31	18	20
	Eschborn	13	4	5
6902	Frankfurt/ Main	76	18	34
	Wiesbaden	55	26	39
6932	AD Mönchhof (A 3)	48	0	0
	Darmstadt	95	20	44
6991	Rüsselsheim	80	7	13
	Mainz	73	6	15
Percentage			33	46

hours exhibiting overlappings are considered more thoroughly. Especially when looking at counting sites with LOS F, it becomes obvious that LOS D cannot be achieved by solely shifting HGV traffic.

Table 5 shows the number of hours among the average weekly patterns exceeding LOS D (separately for counting sites and directions of travel). Besides, these cases are highlighted which require a shift of more than 100 % (or 75 %) of HGV traffic to achieve a sufficient quality of traffic.

The evaluation showed that more than one-third of all cases required a higher shift than the actual HGV traffic volume. Almost half of the cases exhibiting an exceeding of LOS D would require a shift of more than 75 % of HGV traffic to achieve a sufficient LOS. An adequate improvement of traffic flow quality by means of an HGV traffic shift is only possible for counting sites slightly exceeding LOS D.

Aggregation of HGV Traffic Patterns of Different Weekdays

To investigate similarities in the course of the average daily pattern of HGV traffic, these were analysed separately. Two typical patterns were found. On the one hand, there are sections of motorways which show patterns with small changes in the traffic load at certain times. On the other hand, there are sections of motorways which show patterns with a very uniform course of traffic load. However, none of the both types of patterns shows noticeable peak loads.

As an example, Fig. 4 shows the course of a pattern for a typical counting site with small deviations of the traffic loads on different days. As can be seen, the average daily patterns of Tuesday, Wednesday and Thursday are very similar and can easily be clustered. On Mondays and Fridays, the traffic loads are slightly below the loads of the group Tuesday to Thursday. The pattern of Saturday shows a decrease of the traffic loads in evening and night hours, which results from the Sunday truck ban. The traffic load on Sunday increases constantly towards the end of the Sunday ban (8 p.m.).

Figure 5 shows the average daily pattern of the counting site 6902 along the A 66 as an example for daily pattern with a very uniform traffic load. All patterns from Monday to Friday are very similar course without any peak loads. Similar to the pattern described above, Fig. 5 also shows a decrease of traffic loads on Saturday evening and an increase towards Sunday evening. This effect is due to the Sunday ban. However, the number of trucks is very low.

In both cases, the HGV traffic generally has a very smooth course. It can be seen that the load during the day is at a relatively constant level, which is expected to be much higher than the night traffic load.

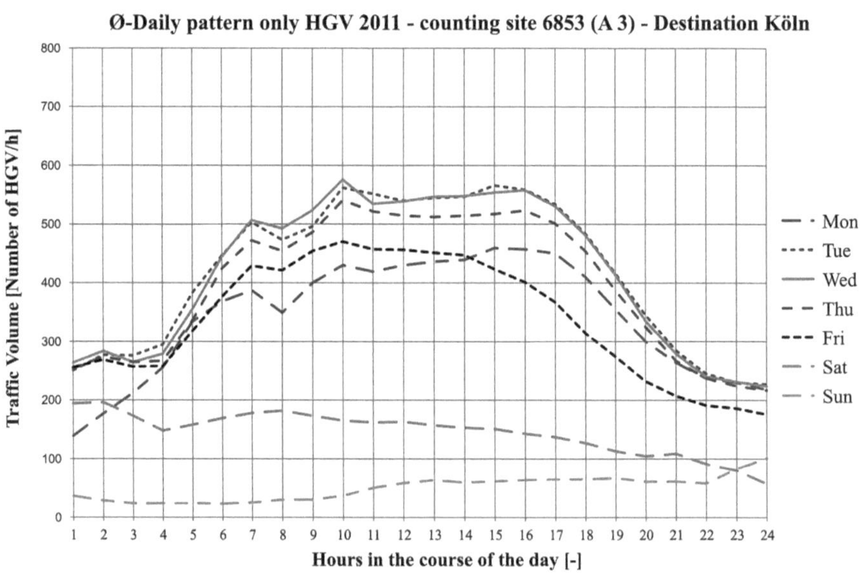

Fig. 4 Average traffic patterns of truck traffic on a specific motorway section with slight differences in volumes on working days

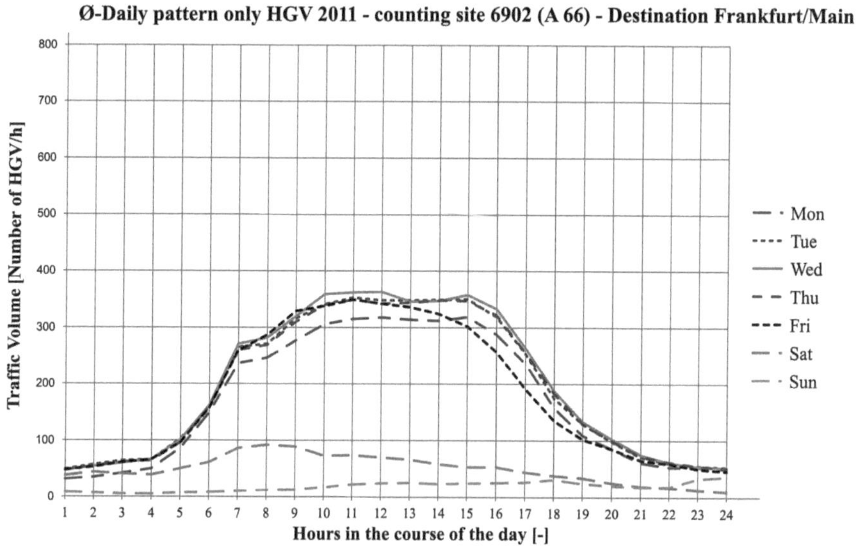

Fig. 5 Average traffic patterns of truck traffic on a specific motorway section with uniform traffic loads on working days

Analysis of the Effects of the Truck Ban During Vacation Time

For the analysis of the effectiveness of the truck ban on Saturdays during the major vacation time, the traffic loads of the affected counting sites during and outside of the validity period of the ban on driving were depicted in one joint diagram. Both counting sites along motorway A 3 (counting site 6853 and 6854) are affected by the truck ban during vacation time. Figure 6 shows the results of counting site 6854 as an example. As can be seen, traffic loads between 0 a.m. and 7 a.m. on a Saturday during the truck ban are higher than those outside. As expected, the number of HGV decreases on routes affected from the truck ban during its validity period between 7 a.m. and 8 p.m. After the end of that period at 8 p.m. the number of HGV increases significantly.

Furthermore, to ensure that the decrease of the traffic load is not an effect of the vacation season, the counting sites of motorways not affected by the truck ban were analysed. Figure 7 shows the analysis of the A 66 (counting site 6902) as an example. It should be noted that there is only a very slight decrease in HGV traffic, certainly due to the vacation season.

Besides the graphical analysis of the results, the change in HGV traffic loads on Saturdays during the truck ban compared to Saturdays outside the ban is summarised additionally in Table 6. It can be seen that counting sites with a ban on driving during vacation time show an average increase of 5 % between 0 a.m. and 7 a.m. The analysis also shows that the ban on driving during vacation time causes an average decrease of about 30 % of the HGV traffic during the affected

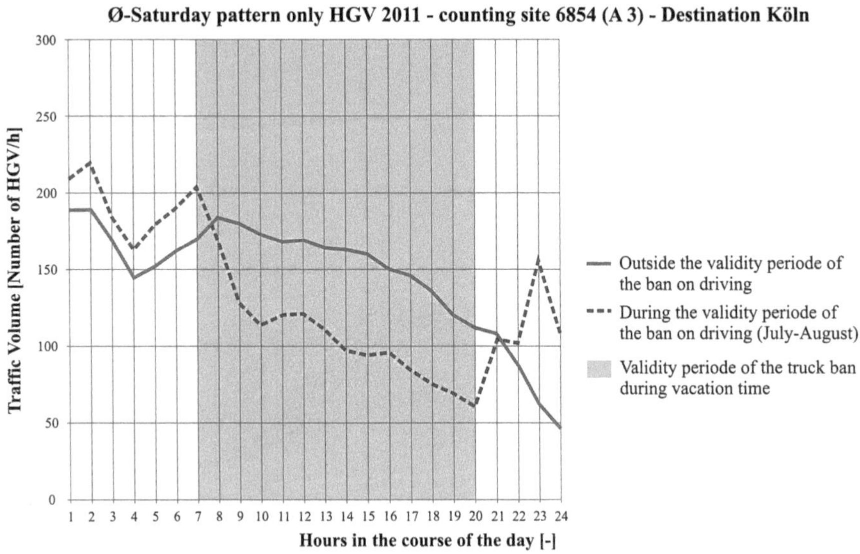

Fig. 6 Comparison of traffic volumes on Saturdays during and outside the truck ban during vacation time (affected counting site)

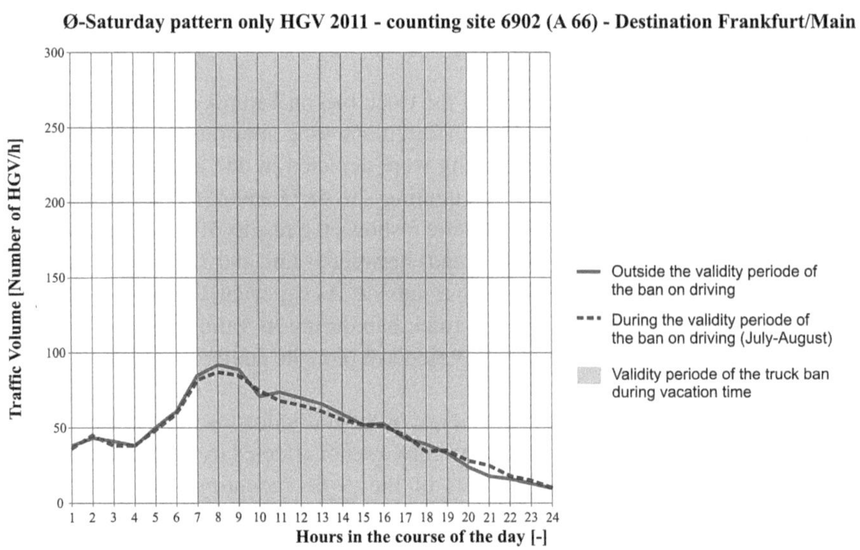

Fig. 7 Comparison of the traffic volumes on a Saturday during and outside the truck ban on driving during vacation time (not affected counting site)

validity period. A higher acceptance would only be feasible with less exceptions and stricter enforcement. After the end of the truck ban at 8 p.m., an average increase of 60 % was observed. On counting sites without a truck ban, there is a

Table 6 Results of the analysis of effects of the truck ban during vacation time

Affected by the truck ban during vacation time?	Time	Range of percental increase or decrease of the truck ban during vacation time (%)	Average percental increase and decrease during the truck ban during vacation time (%)
Yes (Evaluation of a total of 24 patterns on Saturdays)	00:00–06:59 07:00–19:59 20:00–23:59	−5.00 to 14.50 −33.70 to 21.48 50.80 to 69.91	4.9 −28.6 60.0
No (Evaluation of a total of 48 patterns on Saturdays)	00:00–06:59 07:00–19:59 20:00–23:59	−13.60 to 1.82 −16.10 to −0.54 −29.5 to 30.96	−3.1 −9.8 12.1

small decrease of traffic load between 0 a.m. and 8 p.m. and an increase of approximately 12 % after 8 p.m.

Conclusion

Due to the ongoing growth of road transport, the problems of increasing capacity constraints on German roads will intensify.

The data analysis for the estimation of the existing potential for a temporal shift in freight traffic shows that peaks in overall traffic are much more distinctive than the ones in HGV traffic. The constant level of HGV traffic load during the day can be explained with an efficient use of available vehicle and driver resources in freight transport. Contrary to passenger traffic, a long-lasting non-use of the vehicle is hardly acceptable in freight transport due to economic reasons. In addition, many drivers work in the daytime. Thus, shifting truck trips for several hours a day appears not to be a feasible measure to improve traffic quality. Furthermore, in the case of trips for several hours, it cannot be avoided that a part of the trip takes place during the peak hour of the overall traffic. This study only focuses on the local situation. In order to draw conclusions about the effects on the networks, it is necessary to observe a larger research area.

At night, the level of truck load is considerably lower than during the day. Primarily, this can be explained with the timing of production and distribution processes, but partly also with higher labour costs for drivers during the night. In order to achieve an influence of HGV traffic, which contributes to a release of capacity constraints, possible measures in production and logistic should be investigated to enable night transport. The resulting problems of additional nocturnal noise pollution and night work should be taken into account. Further options for traffic reduction and modal traffic shift are to be sought.

It was identified that during the time of peak loads it is not possible to reach a sufficient level of service by only shifting HGV traffic temporally. Since the passenger car traffic mainly contributes to distinct peak loads and stands for a much larger share of total traffic than HGV traffic in the periods of "critical" level of service, it is also required to think about additional measures in passenger car traffic. In the context of considerations to influence the time choice, which should help to improve the traffic flow, cause-dependent shifts of the different vehicle types (depending on the load share of overall traffic) should be generally taken into consideration.

The analysis of the truck ban during vacation time has shown that during its validity both affected counting sites show an average decrease of truck traffic of 30 %.

In Summary This Study Provides the Following Results

- Peak loads in HGV traffic on German motorways are less significant than in other traffic, particularly in passenger car traffic.
- Main reason is that the diurnal course in HGV traffic is considerably more influenced by the efficient use of vehicle and driver resources compared to passenger car traffic.
- Usually, the truck share during peak loads is below 12 % (based on PCU).
- Therefore, the contribution of temporal shift of freight traffic to the avoidance of congestion has to be considered as low.
- Accordingly, approaches for a temporal shift during peak loads should be primarily searched in passenger car traffic, which corresponds with the causation share.
- Possibilities for the temporal shift of truck traffic should primarily be sought in changes of production and logistic processes, which enable night transport. The resulting problems of additional nocturnal noise pollution and night work have to be considered.
- However, measures to avoid traffic, to achieve modal shift and to control traffic are still important in freight traffic.

References

ADAC – Allgemeiner Deutscher Automobil-Club e.V. (2012). *Verkehrsqualität auf deutschen Autobahnen*. München.

Ahrens, G.-A., & Kabitzke, U. (2011). *Zukunft von Mobilität und Verkehr: Auswertungen wissenschaftlicher Grunddaten, Erwartungen und abgeleiteter Perspektiven des Verkehrswesen in Deutschland*. Forschungsbericht FE-Nr.: 96.0957/2010. Dresden: Technische Universität Dresden.

BASt – Bundesanstalt für Straßenwesen. (2011). *Ausführungshinweise zum "Leitfaden zum Arbeitsstellenmanagement auf Autobahnen"*. Bergisch Gladbach.

BASt – Bundesanstalt für Straßenwesen. (2013). Lage der Zählstellen und Erfassungsarten nach TLS. Accessed July 4, 2013, from www.bast.de

Daehre-Komission. (2012). Bericht der Kommission "Zukunft der Verkehrsinfrastrukturfinanzierung" der Verkehrsministerkonferenz.

FGSV – Forschungsgesellschaft für Straßen- und Verkehrswesen. (2012). *Begriffsbestimmungen – Teil: Verkehrsplanung, Straßenentwurf, Straßenbetrieb*. Köln: FGSV-Verlag.

FGSV – Forschungsgesellschaft für Straßen- und Verkehrswesen (Ausgabe 2001, Fassung 2009). *Handbuch für die Bemessung von Straßenverkehrsanlagen* (HBS). Köln: FGSV-Verlag.

HMWVL – Hessisches Ministerium für Wirtschaft, Verkehr und Landesentwicklung. (2013). *Mobilität Hessen – Entwicklungen und Perspektiven*. Wiesbaden.

IVM – Integriertes Verkehrs- und Mobilitätsmanagement Region Frankfurt RheinMain. (2010). *Wirtschaftsverkehr 2030 – Analyse und Prognose des regionalen Wirtschaftsverkehr in der Region Frankfurt RheinMain bis zum Jahr 2030*. Frankfurt am Main.

Kittler, W., & Boltze, M. (2011). Beeinflussung der Zeitwahl von Nutzern des ÖPNV. *Der Nahverkehr, 29*(6), 7–12.

Rühl, F., & Boltze, M. (2016). Freight transport demand management – Influencing the freight transport demand within traffic management. In E. Abele, M. Boltze, & H.-C. Pfohl (Ed.), *Dynamic and seamless integration of production, logistics and traffic. Fundamentals of interdisciplinary decision support* (pp. XXX). Berlin: Springer.

SRU – Sachverständigenrat für Umweltfragen. (2012). *Umweltgutachten 2012 – Verantwortung in einer begrenzten Welt*. Berlin

Strassenverkehrs-Ordnung (German Road Traffic Regulations)

Freight Transport Demand Management: Influencing the Freight Transport Demand Within Traffic Management

Frederik Rühl and Manfred Boltze

Introduction

In the last decades, economic growth, the removal of international trade barriers and the trend towards mass individualisation along with an expanding division of labour in production and logistics have led to a strong growth of freight traffic all over Europe (Tavasszy and Ruijgrok 2013). Despite numerous political initiatives for a modal shift of traffic to railways, freight transport growth in Germany has mainly taken place on the road. While the growth of freight volumes is stagnating, road transport performance has increased rapidly and is still expected to grow (SRU 2012, pp. 137–142). BVU et al. (2014) forecast an increase of road transport performance of about 39 % till 2030. Furthermore, especially for dense urban areas, there are only few alternatives to road freight transport. Growing e-commerce and the liberalisation of the market for courier, express and parcel services mainly contributed to growing traffic volumes (acatech 2012, pp. 20–21). Moreover, current considerations of urban value creation let assume even more freight transport in urban areas in future (Wiegel et al. 2013).

Notwithstanding that road freight transport contributes greatly to preserve our economic prosperity, negative impacts of its strong growth increasingly become apparent. Not only large cities but also small municipalities have to cope with problems such as increasing congestion, air and noise pollution, an increased risk of

F. Rühl (✉) • M. Boltze
Chair of Transport Planning and Traffic Engineering, Technische Universität Darmstadt, Otto-Berndt-Straße 2, 64287 Darmstadt, Germany
e-mail: ruehl@verkehr.tu-darmstadt.de; boltze@verkehr.tu-darmstadt.de

E. Abele et al. (eds.), *Dynamic and Seamless Integration of Production, Logistics and Traffic*, DOI 10.1007/978-3-319-41097-5_9

accidents or infrastructure damage. Likewise, road freight transport growth confronts carriers with challenges such as high stress levels for drivers or decreasing delivery reliability for customers (Ogden 1992, pp. 89–136; SRU 2005, pp. 35–66; Ahrens et al. 2010; Taniguchi et al. 2012, pp. 40–45).

There have been numerous initiatives all over Europe aiming for curbing freight transport's negative impacts by means of increasing the efficiency of delivery processes or promoting cooperation between public authorities and enterprises. Although their efficiency has been proven, these initiatives are the exception rather than the rule yet. To protect residents, decision-makers in transport authorities usually still counteract the increasing problems by enforcing restrictive measures such as night-time truck bans (Dablanc 2007; Ballantyne et al. 2013; Lindholm and Browne 2013). Indeed, such restrictive measures achieve their objectives, e.g. by fulfilling the goals for emission reduction. However, there is still no consideration of impacts on freight transport and the closely related concepts of production, logistics and retail of such restrictive measures (Stathopoulos et al. 2012). For instance, the introduction of low emission zones in German cities slightly reduced the emission levels of particulate matter and nitric oxides (Boltze et al. 2014). But still, no attention is paid to the impacts of environmental zones on business processes. For example, it is possible that small companies are forced to modernise their fleet while bigger competitors may still employ older vehicles in regions without environmental zone (Anderson et al. 2005). Reasons why decision-makers in transport authorities neglect measures' impacts on production and logistics are manifold. On the one hand, transport authorities often lack required expert knowledge or specific data (Ballantyne et al. 2013). On the other hand, absent evaluations and hesitant exchange between municipalities worsen learning from failed projects (Lindholm and Blinge 2014).

Basically, also enterprises may help to stem the mentioned problems of freight transport growth. Definitely, they may simultaneously gain a direct or indirect financial benefit by initiating and implementing measures such as joined deliveries, efficient driving or a shift of transports to off-peak hours (Anderson et al. 2005). But up to now, they usually rarely consider impacts of their decisions on traffic and transport. Carriers mainly orientate operations to their customer's needs (Dablanc 2007). Again, shippers optimise own processes under consideration of internal financial and logistical requirements or arrange processes with their supply chain partners, respectively. Transport costs with their marginal ratio to total costs usually only play a minor part in such considerations (Anderson et al. 2005).

In summary, both decision-makers in traffic and transport and decision-makers in supply chains are practising an insufficient consideration of the needs of the counterpart. Though, to meet the challenges mentioned above, stakeholders have to cooperate. Furthermore, present habits have to change, i.e. public authorities have to consider impacts of their measures on business processes and enterprises have to account for the induced traffic of their business. Thus, freight transport demand has to be influenced. While mobility management has already been used for influencing demand in passenger traffic since two decades, traffic management still lacks in a concept for influencing the freight transport demand.

Accordingly, this chapter presents the concept of freight transport demand management as part of traffic management based on Boltze et al. (2012) and Boltze (2013). By analogy with mobility management for passenger traffic, freight transport demand management aims at influencing the demand for freight transport. For this purpose, the development of freight transport demand and relevant stakeholders are discussed in the first step. Subsequently, definitions and principles of freight transport demand management are introduced and existing approaches are described. The following chapter outlines the variety of measures that can be used for influencing the freight transport demand. The chapter closes with a conclusion and an outlook which determines the need for action.

Relevant Stakeholders for Influencing the Freight Transport Demand

According to Manheim (1979), the emergence of freight transport can be explained with the interaction of the transport system with the entirety of political, economic and societal activities as well as the flows of goods themselves. He states that both the physical elements of the transport system and the decisions in the activity system have an influence on the realisation of transports. Furthermore, realised and expected flows of goods have a bearing on decisions in the transport and activity system.

As the locations of supply and demand for a good usually differ, a movement of goods on the traffic infrastructure is required. Freight transport is constituted by the entirety of all goods movements. Thus, freight transport demand is a derived demand. Accordingly, influencing the freight transport demand requires the identification of the driving forces, which induce flows of goods and, thus, freight transport (Anderson et al. 2005). On the one hand, these driving forces are stakeholders whose activities directly induce freight transport. On the other hand, also stakeholders who make decisions changing the conditions of the freight transport system influence freight transport. Thus, to influence freight transport demand, it is necessary to understand the specific goals which underlie the decision-making of freight transport's relevant stakeholders.

According to Östlund et al. (2002), freight transport demand derives from the interaction of the decision-making levels productions, retail, warehousing and transport which are interrelated via supply and demand. Consequences of that interaction are the observable transports on the traffic infrastructure (freight transport). Rühl and Ottemöller (2013) seize on that framework and identify the markets for goods, transports and traffic—that is the interfaces between the freight transport-relevant systems consumption/intermediates, production/retail, logistics and traffic defined by supply and demand—as leverage points for freight transport-related measures (see Fig. 1). Thus, changing the behaviour of the relevant stakeholders for these markets allows for influencing freight transport demand.

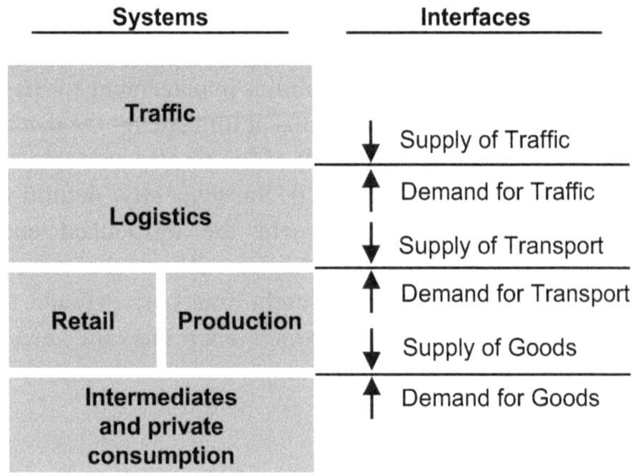

Fig. 1 The markets of freight transport demand (Rühl and Ottemöller 2013)

Literature gives various definitions of relevant stakeholders in freight transport with different levels of detail. Muñuzuri et al. (2005) only consider carriers, receivers and public authorities in their overview on measures for sustainable freight transport since these stakeholders can implement such measures. Russo and Comi (2011) as well as Taniguchi et al. (2012) also incorporate shippers, private consumers and inhabitants as further stakeholders with an influence on freight transport. The broad definition of Ballantyne et al. (2013) differentiates stakeholders with a direct influence on freight transport (inhabitants, shipper, receiver, carrier, public authorities) as well as stakeholders whose activities indirectly affect freight transport (inter alia trade associations, public transport operators).

Referring to Ballantyne et al. (2013), in the following, the focus lies on the stakeholders with a direct impact on freight transport demand. On the one hand, those are receivers, shippers and carriers as the elements of the transport chain. On the other hand, also private infrastructure operators and public authorities impact freight transport since they provide infrastructure and determine usage rules. These stakeholders can be assigned to the three markets which determine freight transport demand (see Fig. 2).

The *receiver* determines general transport conditions with his decision on purchasing a certain good together with the decisions regarding purchasing source and date. His motivation is a procurement of the favoured good at minimal costs meeting his specific requirements.

The *shipper* as goods supplier and costumer of the carrier aims at satisfying the customer needs by delivering the good at minimal costs meeting the requirements. For this purpose, he impinges on the delivery conditions for the carrier.

The *carrier* equally strives for satisfying the needs of receiver and shipper with a high-quality transport at minimal costs. To bundle transports he makes use of his logistical infrastructure (warehouses, etc.).

The *infrastructure operator* provides infrastructure and determines usage rules under consideration of statutory requirements of the public authorities. He aims at insuring an efficient traffic system at minimal costs.

<table>
<tr><th colspan="2">Markets constituting Freight Transport Demand</th><th>Stakeholder</th><th>Objective of Stakeholder</th></tr>
<tr><td rowspan="2">Goods market</td><td>Demand of goods</td><td>Receiver</td><td>Procurement of favoured good at minimal costs meeting the requirements</td></tr>
<tr><td>Supply of goods</td><td rowspan="2">Shipper</td><td rowspan="2">Satisfying customer needs by delivering the good at minimal costs meeting the requirements</td></tr>
<tr><td rowspan="2">Transport market</td><td>Demand of transport</td></tr>
<tr><td>Supply of transport</td><td rowspan="2">Carrier</td><td rowspan="2">Satisfying customer needs by transporting the good at minimal costs meeting the requirements</td></tr>
<tr><td rowspan="3">Traffic market</td><td>Demand of traffic</td></tr>
<tr><td rowspan="2">Supply of traffic</td><td>Infrastructure operator</td><td>Provision of an efficient traffic system at minimal costs</td></tr>
<tr><td>Public authorities</td><td>Provision of an efficient traffic system considering economic, ecological and social requirements</td></tr>
</table>

Fig. 2 Relevant stakeholders and their objective in the three markets constituting freight transport demand

The *public authorities* provide infrastructure and determine usage rules. They strive at insuring an efficient traffic system considering economic, ecological and social requirements. This demands a weighting to the respective interests of all stakeholders.

Transport Demand Management Within Traffic Management

Mobility Management to Influence the Demand in Passenger Traffic

The traffic supply oftentimes cannot be extended in accordance with the increasing traffic demand. As a consequence, mobility management has found its way into European transport research at the beginning of the 1990s being developed from the US-American transportation demand management. Meanwhile, mobility management may be rated as established in research and is also increasingly used in practice (Bruns and Langweg 2010; Reutter 2014). The following excursus highlights the concept of mobility management looking closely at its principles. Furthermore, the lacking consideration of freight transport within that concept is pointed out.

Despite various slightly different definitions (ILS/ISB 2000; UBA 2001; Fiedler 2002; Bruns and Langweg 2010; FGSV 2012; EPOMM 2013), it can be stated that mobility management aims at influencing traffic behaviour and modal choice in passenger transport to optimise traffic's positive and negative impacts. For this purpose, target group-specific bundles of mostly "soft measures" (measures for information, communication, organisation and coordination) but also regulatory and pricing measures are deployed. The scope of measures reaches from influencing the land use via the flexibilisation and staggering of working hours to fiscal measures such as congestion charging and parking management. For a comprehensive list of mobility management measures, see for example EPOMM (2013). These measures address the individual traveller in order to change his attitudes concerning mobility and the use of different means of transport. As a result of this, the individual shall independently strive for a more efficient and sustainable use of existing traffic systems (Kemming 2009).

The public authorities as provider and initiator of mobility management—for instance by means of the appointment of a mobility manager (Lucietti 2012)—address homogenous groups, i.e. groups of people with similar travel behaviour, via comprehensive and target group-specific bundles of measures. Examples for such fields of action are municipal, location-related or corporate mobility management (Kemming 2009; Blees 2010).

There is broad agreement about mobility management's contribution to sustainable transport. However, particularly long-time impacts of mobility management, that is impacts on travel behaviour, are difficult to observe or quantify (Gertz et al. 2012). Thus, politicians are still reluctant with its implementation although mobility management measures are a cost-efficient alternative to improve the traffic situation compared to expensive infrastructure investments (Ahrens and Hubrich 2012).

To which extent mobility management addresses and influences the freight transport demand has to be challenged. Indeed, former definitions of mobility management comprised freight transport. Fiedler (2002) as a pioneer of mobility management in Germany still addressed freight transport in his mobility management approach. Likewise, mobility management influences passenger and freight traffic according to the definition by ILS/ISB (2000). Nevertheless, newer definitions increasingly highlight the focus on passenger transport (see e.g. Bruns and Langweg 2010; FGSV 2012; EPOMM 2013). Also, a brief look at implemented measures and fields of action mentioned above suggests that freight transport is not part of mobility management.

Notwithstanding the above, due to the ambiguous definitions there are still examples which dedicate freight transport to the concept of mobility management. Within the EU project C-LIEGE, measures to influence the freight transport demand were collected and assigned to mobility management. Likewise, the implementation of a city logistics manager as counterpart to the mobility manager for passenger traffic focusing on the planning and implementation of freight transport measures is demanded (Lucietti 2012; Torentellé et al. 2012). Also according to the Online TDM Encyclopaedia of VTPI (2015), influencing the freight transport demand is part of transportation demand management, also called mobility management. The entry to the term "freight transport management" presents various regulatory and pricing but also innovative measures such as transport consultancy for enterprises as strategies for more efficient freight transport. But again, a look at implemented transport demand management measures (see Litman 2013) shows the strong focus on passenger transport.

In summary, it can be concluded that mobility management has been proven to be a successful instrument to change travel behaviours. However, freight transport is conceptually as well as in practice explicitly not in the focus of mobility management. Thus, so far, traffic management lacks a concept for influencing freight transport, being equal to mobility management.

Existing Approaches for Influencing the Freight Transport Demand

Since the mid-1990s, freight transport increasingly took centre stage of traffic policy due to problems going along with it; that is nearly contemporaneous to the arising of mobility management. As written above, that concept for influencing demand in passenger transport neglects freight transport. Basically, the idea to influence freight transport is nothing new. However, there are only few approaches for the development and implementation of a concept for freight transport demand management analogous to mobility management.

Considering the different conditions for passenger and freight traffic, Binnenbruck et al. (1998) motivate to develop an adjusted mobility management for freight transport. The suggested measures mainly comprise incentivisation and consulting. Taniguchi and Nemoto (2002) highlight the need for freight transport demand management to cope with the increasing problems caused by freight transport and overview potential measures. Their broader scope of measures also considers regulatory, pricing and technological measures. Bretzke and Barkawi (2012) doubt the effectiveness of freight transport demand management measures to cope with the increasing problems related to freight transport. Nonetheless, they suggest their implementation. Lucietti (2012) and Torentellé et al. (2012) go one step further and present an extensive catalogue of soft measures for influencing freight transport demand as well as a concept for the introduction of a city logistics manager as mentioned above. VTPI (2015) lists regulatory and pricing measures but also various innovative soft measures such as transport consultancy for enterprises as strategies for more efficient freight transport. The broadest approach is introduced by Taniguchi et al. (2012). They recommend a stepwise approach to develop sustainable "road freight transport management" which allows for stakeholder involvement within planning, implementation and evaluation phase. To cope with freight transport's negative impacts, they suggest bundles of infrastructure and soft measures.

The arising of that new field of freight transport research went along with the increasing development of measures to contribute to sustainable transport in response to the increasing freight transport problems at that time. Initially, many projects focused on measures contributing to more efficient freight processes themselves. At the beginning, it was mainly sought for possibilities to bundle transports. Famous examples for this are urban consolidation centres which were founded especially in Germany in large numbers by the end of 1990 (e.g. Köhler 1999). However, due to their inefficiency, many of them have already been closed again. In the following years, several EU projects (i.e. BESTUFS, SUGAR, CIVITAS) investigated topics such as the use of alternative vehicle technologies or innovative delivery concepts for the last mile (Allen et al. 2008; Dablanc 2011; van Rooijen and Quak 2014). In the last few years, research also focused on cooperation between public authorities and the relevant stakeholders of freight transport. Such a strong and continuous cooperation is seen as crucial success factor for sustainable freight transport, since it enables the development of measures with general consensus and, thus, sustainable freight transport strategies (Allen et al. 2010; Lindholm and Browne 2013). Accordingly, there are numerous works presenting lists of measures. Besides the above-mentioned catalogues of best practices from EU projects also Muñuzuri et al. (2005), Quak (2008), Russo and Comi (2011) and Dablanc et al. (2014) show comprehensive lists of measures.

Requirements for a Concept to Influence Freight Transport Demand Within Traffic Management

In summary, it can be stated that mobility management is successfully used for influencing the demand in passenger traffic. But freight transport is explicitly not in the focus of that concept. There are hints and first approaches for the development of a demand management concept for freight transport. However, no comprehensive concept similar to mobility management has been presented yet. Thus, until now, traffic management lacks in a concept for influencing the freight transport demand, being equal to mobility management.

Considering mobility management, such an approach for influencing the freight transport demand within traffic management needs to fulfil the following requirements:

- It has to aim at an integrated optimisation of freight transport's impacts considering impacts on traffic and the environment as well as business and economy.
- It has to address groups with homogenous behaviour based on a careful definition of different target groups.

- It has to provide strategies for short-term and long-term freight transport demand management in terms of target group-specific bundles of measures.
- It has to promote the insitutionalisation of freight transport by means of its consideration in transport planning instruments as well as involvement of the relevant stakeholders into planning and implementation of freight transport-related measures.

As first steps for the development, the following chapters deal with the integration of freight transport demand management into traffic management and the derivation of principles for the concept as well as the discussion of suitable measures.

The Concept of Freight Transport Demand Management

Definitions

As the first step, the development of freight transport demand management needs a clear distinction from mobility management by means of clear definitions and integration into traffic management framework.

> *Traffic management* influences the supply of traffic and transport systems as well as the demand for travel and transport through a bundle of measures with the aim to optimise the positive and negative impacts of traffic and transport (FGSV 2012).

Traffic management aims at an optimisation of the impacts of traffic by employing instruments for influencing the traffic supply as well as the traffic demand. Traffic supply management comprises the provision and operation of traffic systems. Insofar as the traffic supply, in particular the road infrastructure, is shared by passenger and freight traffic, measures which influence the traffic supply have to be seen in a common context. Furthermore, supply measures are hard to dedicate to passenger or freight traffic because of strong interdependencies between the two of them. By contrast, traffic demand management may be divided into influencing the demand for passenger transport (known as *mobility management*) and influencing the demand for transport of goods, called *freight transport demand management* (see Fig. 3).

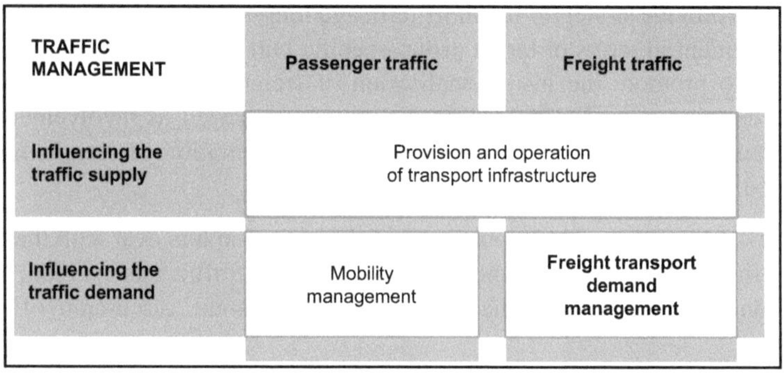

Fig. 3 Freight transport demand management as part of traffic management (Boltze 2013)

Hence, both concepts can be defined analogously as follows:

> *Mobility management* influences the demand for passenger transport by implementing bundles of measures in order to optimise the positive and negative impacts of traffic and transport.
> *Freight transport demand management* influences the demand for freight transport by implementing bundles of measures in order to optimise the positive and negative impacts of traffic and transport.

The need to distinguish between two concepts for passenger and freight transport may be explained with the different conditions for passenger and freight traffic. The origination processes of changes of location differ for people and goods. Thus, the possibilities to influence the demand differ as well. To influence the demand in passenger traffic, it is necessary to impact on the individual's demand for mobility, which arises from the interaction of its personal transport desire, transport requirement and transport ability. For example, factors such as economic status, age, car availability or city characteristics have influence on passenger traffic demand. By contrast, a concept for freight transport demand management has to account for the even more complex interactions between goods market, transport market and traffic market, i.e. the requirements and objectives of different stakeholders have to be addressed (see chapter "Interdisciplinary Decisions in Production, Logistics, and Traffic and Transport: Measures for Overcoming Barriers in Interdisciplinary Decision Making").

Principles of Freight Transport Demand Management

Because of mobility management's success to change traffic behaviour in passenger transport, it seems obvious to develop and implement a concept of freight transport demand management based on the principles of mobility management. In the following, principles for freight transport demand management are developed based on the mobility management principles elaborating issues regarding its objectives, suitable measures, its fields of action and its institutionalisation.

- *Objectives*: Freight transport demand management aims at the optimisation of freight transport's impacts. In terms of an integrative planning approach, that optimisation does not only consider impacts on traffic and the environment, but also impacts on enterprises and resulting economic impacts for the society. Thus, freight transport demand management provides a system of material and immaterial incentives across all transport modes to enable influencing the freight transport demand by changing transport behaviour.
- *Measures*: Similar to mobility management, bundles of measures are implemented which neither demand any structural changes of the existing infrastructure nor a construction of new infrastructure. Amongst others, these are measures like the issuing of guidelines and regulations as well as pricing measures to influence the modal choice (that is both the choice of transport mode and the choice of the type of vehicle), time choice or route choice. Likewise, soft measures like recommendations as well as measures for information, communication, organisation and coordination are applied. These measures aim for initiating changes of production and logistics concepts, which go along with relief for traffic and the affected people.
- *Fields of action*: Freight transport demand management focuses on target groups with homogenous behaviour to be able to consider different transport needs of enterprises and ensure high acceptance of measures. Possible target groups are receivers, shippers and carriers as the main actors of freight transport. Moreover, sector-specific and location-specific fields of action may be developed.
- *Institutionalisation*: Freight transport demand management is provided and initiated by the public authorities to promote the general public interests. Following the idea of integrated traffic planning, relevant stakeholders are involved into planning and implementation of freight transport demand management measures as early as possible. That concerns a federal and cross-organisational cooperation of public decision-makers as well as a strong and long-lasting cooperation with receivers, shippers, carriers, infrastructure operators and other social stakeholders to ensure the development of sustainable solutions. Furthermore, measures of freight transport demand management are coordinatedly implemented into the different transport planning instrument, such as traffic development plan.

As next step to further develop the concept of freight transport demand management, the following chapter outlines the variety of measures that can be used for influencing the freight transport demand by assigning them to the hierarchical levels of traffic management and the stakeholders addressed.

Measures of Freight Transport Demand Management

Categorisation of Measures

To class the variety of freight transport measures with the concept of traffic management and to deduce fields of actions, suitable measures identified in the literature are initially clustered into groups of similar measures (e.g. fiscal instruments or consulting/training). Subsequently, these are classified into the hierarchical levels of traffic management [traffic reduction, traffic diversion (spatial, temporal, and modal), traffic control] as defined in Boltze (1996). Accordingly, it can be distinguished into

- *Measures for traffic reduction*, which aim for reducing the traffic performance (veh*km).
- *Measures for spatial shift of traffic*, which aim for influencing the source and destination choice of freight transport to reduce the transport performance (tkm).
- *Measures for temporal shift of traffic*, which aim for changing the temporal dispersion of the traffic volume (veh/h).
- *Measures for modal shift of traffic*, which aim for changing the modal split (%).
- *Measures for traffic control*, which aim for establishing an efficient handling of traffic under consideration of various parameters such as traffic flow or environmental indicators.

In addition, measures can be differentiated regarding the decision-maker who is able to implement them. While regulatory or pricing measures (e.g. truck bans or HGV tolls) are in the direct area of responsibility of the public authorities, most of changes of production and logistics processes highly desirable for sustainable freight transport such as shifts of transport to off-peak hours are solely realisable by receivers and shippers or carriers. The public authorities may only impinge on them with incentive measures or the offer of consulting and training. The classification resulting from that is as follows:

- *Measures influencing infrastructure usage rules*, which are mainly regulatory and pricing measures.
- *Measures influencing carrier behaviour*, which are mainly soft measures.
- *Measures influencing receiver/shipper behaviour*, which are mainly soft measures.

Furthermore, the public authorities may implement supportive measures to foster communication between the freight transport stakeholders, to extend the general knowledge about the freight transport system and promote the acceptance of freight transport in the public perception.

As shown in Table 1, regarding the hierarchical levels of traffic it becomes obvious that already many measures for traffic reduction are in use. By contrast, the table displays a lack of available measures for spatial and temporal shifts of traffic which were required for a better use of the existing infrastructure. Moreover, while

Table 1 Measures of freight transport demand management

Hierarchical level of traffic management	Measures influencing Infrastructure usage rules	Carrier behaviour	Receiver and shipper behaviour
Traffic reduction Reduction of traffic performance (veh*km)	+	+	+
Spatial shift of traffic Reduction of transport performance (tkm)	–	–	(+)
Temporal shift of traffic Change of temporal dispersion of traffic volume (veh/h)	+	–	(+)
Modal shift of traffic Change of modal split (%)	+	(+)	(+)
Traffic control Efficient handling of traffic (traffic flow, emissions, ...)	+	(+)	–
Supportive measures	(+)		

many implemented measures aim for influencing infrastructure usage rules, the carrier or receiver/shipper behaviour is still not insufficiently influenced with appropriate soft measures. This approves the proposition stated above that decision-makers in transport authorities usually rather draw on restrictive than on soft measures. Accordingly, it can be stated that there is a particular need for research for the development and trial of soft freight transport demand management measures.

In the following, based on the hierarchical levels of traffic management, the different measures are briefly discussed by describing their objective and current status of implementation.

Measures for Traffic Reduction

Traffic planners have several measures for traffic reduction on hand. Amongst others, they can use *planning instruments* to influence the trip generation. An established measure is the designation of easily accessible areas, e.g. with direct connection to the motorway network, for freight-intensive business of receivers, shippers or carriers. By this means, trips through living areas and, thus, impairments for residents may be limited (see Taniguchi et al. 2012). Furthermore, legally binding regulations regarding parking spaces in land-use plans prevent traffic disturbance due to delivery vehicles (see Dablanc et al. 2014). Likewise, cities may use their transport development plans to influence freight transport by defining corresponding targets and strategies for their traffic policy. However, freight transport is usually still disregarded in these plans. But the integrative development process of such plans allows for the involvement of all relevant stakeholders. Hence, there is the possibility to integrate sustainable freight transport strategies with general consensus as principles for the planning and implementation of freight transport measures (see Lucietti 2012). Not yet disseminated are statutory rules demanding logistics concepts for construction sites to limit the induced traffic, e.g. with bundling of transports. Such regulatory requirements for planning permission shall motivate building companies to consider impacts on traffic within their optimisation of logistical and operational processes on the construction site (see Browne et al. 2012). Another innovative idea is the *limitation of the traffic volume* by means of mobility budgets for receivers and shippers in urban areas. These budgets only allow for a limited amount of own trips as well as trips by service providers free of charge per year. Additional trips require charged certificates (see Torentellé et al. 2012). On the national level, *fiscal instruments* such as mineral oil tax or motor vehicle tax are tried and trusted to influence transport costs (see UBA 2009).

A well-established instrument is the *promotion of carrier cooperation*. It strives for reducing traffic volume by means of transport bundling or shared use of logistical infrastructure. Amongst others, these are initiatives for joint deliveries or the operation of urban and suburban consolidation centres (see Quak 2008; Dablanc et al. 2014). Likewise, *promotion of cooperation between carrier and receivers/shippers* is an approved measure. For instance, online freight exchanges allow for better coordination between the stakeholders and then for increased vehicle load factors. Shared freight reception points reduce the number of carrier's trips, too (see Quak 2008; Lucietti 2012). Furthermore, *certification of carriers* according to sustainability standards as done with the Freight Operator Recognition Scheme in the United Kingdom contributes to increase carrier's process efficiency. Similarly, the certificate guarantees the customers a high transport quality, so that the carrier gains competitive advantage (see Browne et al. 2012).

Programmes for (voluntary) *certification of receivers and shippers* with high publicity potential shall encourage them to contribute to sustainable transport with adherence to jointly developed standards. Incentives may either be public perception of the awarded certificate or benefits such as flexibilisation of delivery windows. Corresponding programmes have been implemented in London and the Netherlands (see Dablanc et al. 2014). Moreover, *consulting services for receivers and carriers* by public authorities are a possible measure. In the context of such an offer, companies may be informed about their opportunities to contribute to traffic reduction, e.g. change of product design and packaging for increased load factors or adjustment of delivery strategies (VTPI 2015).

Measures for Spatial Shift of Traffic

Contrary to the numerous measures for traffic reduction, there are only few measures for influencing the destination choice of traffic, since that is significantly determined by the receiver with his sourcing decision. Thus, collaboration between enterprises located in the same region has to be fostered with the *promotion of corporate cooperation* in the form of cluster initiatives or the organisation of trade fairs so that goods are rather sourced regionally than globally. However, stronger regional cooperation may lead to increased exports. Thus, it is not possible to estimate the impact of clustering on traffic (see UBA 2009). Besides, consulting services and certification as already implemented for traffic reduction may contribute to a change of sourcing behaviour. For instance, gratuitous calculation of the carbon footprint or awarding certificates for "traffic-friendly receivers" makes for sensitising enterprises for the consideration of traffic by illustrating them the impacts of their sourcing decisions on traffic.

Measures for Temporal Shift of Traffic

Influencing the time choice to avoid congestion and limit environmental pollution is important for both passenger transport (see Kittler and Boltze 2011) and freight transport (see Menges and Boltze 2016). For this purpose, particularly *temporal vehicle bans* come into use. That comprises temporal truck bans such as the German holiday ban on trucks, night-time truck bans being implemented in many German cities due to environmental reasons or the daytime truck bans in Asian megacities being implemented due to capacity constraints (see Son 2006). Another appropriate instrument is the flexibilisation of existing or the implementation of delivery time windows. Delivery windows are already in use for many pedestrian zones to shift deliveries to off-peak hours (see Quak and de Koster 2009). Similar objectives shall be achieved with *time-differentiated road pricing* concepts. Such congestion charging schemes as implemented in London (Givoni 2012) or Stockholm (Börjesson et al. 2012) induce shift of transports with increased charges during peak hours.

Financial support may motivate receivers and shippers to *temporal shifts of deliveries to off-peak hours*. A corresponding pilot project was launched in New York City. Urban receivers were financially supported for accepting night-time deliveries, that is the reception of goods outside operating hours. The study indicated promising shifting potentials (see Holguín-Veras et al. 2011). Moreover, *consulting services* concerning the change of production processes and warehousing concepts may lead to temporal shifts. The business decision on producing in two or three shifts has significant impacts on the temporal dispersion of passenger and freight traffic. Likewise, decisions on the number of inbound and outbound warehouses combined with the choice of logistics concepts such as just in time or just in sequence significantly affect the temporal traffic dispersion.

Measures for Modal Shift of Traffic

Influencing the modal choice in freight transport, in particular shifting freight transports to railways, is a long-standing objective of transport policy. Furthermore, technological progress is fostered for a long time with the *setting of general standards*. For example, emission or safety standards ensure a frequent fleet modernisation (see Dablanc et al. 2014). In addition, *vehicle-dependent driving restrictions* with regard to vehicle type, weight, size, emission class or load factor are very common. The achieved impacts depend on the measure design. While emission class-dependent restrictions such as environmental zone primarily target fleet modernisations, vehicle size-depending restrictions are, for example, enacted for delivery vehicles in historic city centres with narrow streets (see Quak 2008; Russo and Comi 2011). Further, *vehicle-dependent road pricing* schemes aim for fleet modernisation and possibly a change of modal split. For example, the introduction of the distance-based HGV toll system in Germany (whose amount depends, amongst other parameters, on the emission class) has led to truck fleet modernisation, but not to a significant shift of freight transport to railways (see Rühl et al. 2013).

Fleet modernisation may also be fostered with financial support by means of *vehicle support programmes*. These focus on either vehicle modification or fleet modernisation. Additionally, promotion of alternative transport vehicles such as electric vehicles or trams contributes to the modal shift (see Dablanc et al. 2014).

For a long time, receivers and shippers are motivated to use sustainable transport vehicles. However, these *financial promotion programmes for infrastructure*, e.g. for the construction of or recommissioning of existing railway sidings (see BMVBS 2010), are of little success (see Dablanc et al. 2014). *Consulting and training services* to promote the consideration of external costs within the decision-making on the modal choice (see VTPI 2015) are rather innovative. One of its first implementations is the Green Deal programme in the Netherlands (see Government of the Netherlands 2015).

Measures for Traffic Control

Basically, measures for influencing route choice and driving behaviour are part of traffic supply management. However, they also influence the traffic demand. These include *spatial driving restrictions* such as truck bans for transit traffic (see Castro and Delos Reyes 2010), *parking regulations* differentiated by time and space [e.g. loading/unloading zones (see Dezi et al. 2010)] as well as *spatially differentiated road pricing schemes* (see Quak 2008). A good example for that are HGV toll reductions for less noise-sensitive motorways in Japan, by which means significant relief for noise-sensitive alternative routes could be achieved. Carriers' *route optimisation* is improved with the implementation of truck routes (see Miltner

2011) or information and guidance systems (see Dablanc et al. 2014). For *traffic flow smoothing*, speed limits (see UBA 2009) come into use amongst other things. Still very rare are *truck prioritisations* at signalised intersections (see Lucietti 2012) or the designation of truck-only lanes (see Fischer et al. 2003) although such measures may reduce the number of stops, emissions and journey times.

Efficient driving behaviour is enhanced with *driver trainings* for carriers (see Muñuzuri et al. 2005). Likewise, the *promotion of new technologies* increases freight transport efficiency. For instance, the introduction of driver assistance systems like vehicle-interval warning systems blind spot detection systems particularly increase traffic safety.

Supportive Measures

Meanwhile, several initiatives have been implemented to *improve communication and cooperation between the stakeholders of freight transport*. For example, knowledge platforms with information about freight transport measures' regulations, contact persons, etc. [such as the Freight Information Portal of Transport for London (Dablanc 2011)] or the appointment of a contact person for freight transport issues [e.g. a City Logistics Manager (Lucietti 2012)] ease enterprises' contacting of the public authorities. Strong cooperation between the relevant stakeholders helps developing sustainable freight transport policies. For example, "round tables" enable a participative planning and implementation process of freight transport measures (Allen et al. 2010; Taniguchi et al. 2012).

Image campaigns for sustainable freight transport as well as concepts for "green logistics" (see UBA 2009) may raise enterprises' awareness for sustainable transport. Likewise, image campaigns are suited for changing the negative image of freight transport in public perception and, thus, to advance the acceptance of its negative concomitants.

Conclusions

In the future, influencing the freight transport demand will become more important in light of the increasing problems related to freight transport. Mobility management is successfully used for influencing the demand in passenger traffic. But freight transport is explicitly not in the focus of that concept. Furthermore, there is no comprehensive approach for freight transport demand management yet. Thus, until now, traffic management lacks in a concept for influencing the freight transport demand, being equal to mobility management.

Correspondingly, this chapter presents a concept of freight transport demand management within traffic management. The principles for that concept have been derived from mobility management. Furthermore, a broad overview about freight transport demand management measures has been given. Next necessary steps for the future development of the concept are the deduction of fields of action with corresponding bundles of measures and development of possibilities for the institutionalisation of freight transport demand management.

While the need for the application of bundles of measures to influence freight transport demand is pointed out in the literature, transport planners usually still draw on single (restrictive) measures. The deduction of sector-specific and region-specific fields of action of freight transport demand management for receivers, shippers or carriers as well as the development of target group-specific bundles of measures helps transport planners with decision-making. It can be assumed that the scope of measures to influence urban retailers is different to the one for influencing producing companies in rural areas. Due to the assignment of measures to the hierarchical levels of traffic management and the stakeholders addressed with them, the catalogue of measures presented in this chapter can be subsequently used to deduct different fields of action for freight transport demand management with corresponding target group-specific bundles of measures.

"Round tables" or the implementation of a city logistics manager are promising approaches for the institutionalisation of freight transport demand management. Based on the experience gained from these measures, it needs development and trial of further participative concepts. We expect that intensive collaboration between the relevant stakeholders of freight transport and the public authorities can make a significant contribution to sustainable transport. The keys to success of such a concept are the willingness and ability of the stakeholders for cooperation and a design of bundles of measures with advantages for all parties. In addition, freight transport has to be implemented into the transport planning instruments which serve as a basis for decision-making in transport policy.

Furthermore, there is need for research regarding the (further) development and trial of individual freight transport demand management measures. Multifarious measures have been developed which can be assigned to freight transport demand management. While the big variety of measures aiming for more efficient freight processes is already well investigated, especially innovative measures such as consulting services by the public authorities need further investigation. This comprises the spreading of best practices as well as the advancement of first approaches for consulting and training services or certifications of receivers, shippers and carriers by the public authorities to contribute to sustainable freight transport with changes of processes in production and logistics.

As with mobility management, long-term impacts of freight transport demand management measures are still not investigated. Hence, by means of in-depth impact analysis, the impacts of hard and soft freight transport demand management measures on business decisions and induced changes of traffic, i.e. changes of modal split as well as temporal and spatial dispersion of freight transport, have to be examined. Main challenge will be to find a macroscopic scope which considers the interests of all relevant stakeholders of freight transport. The characteristics and requirements of enterprises seem to be sector specific and individual so that generally detailed individual considerations seem necessary. The development and implementation of a performance measurement system for freight transport demand management would enable the evaluation of implemented measures. Based on that, freight transport strategies could be evaluated according to general standards and continuously improved in the sense of quality management.

References

acatech. (2012). Menschen und Güter bewegen. In *Integrative Entwicklung von Mobilität und Logistik für mehr Lebensqualität und Wohlstand*. Heidelberg: Springer Vieweg.

Ahrens, A., Baum, H., Beckmann, K., Boltze, M., Eisenkopf, A., Fricke, H., et al. (2010). Internalisierung externer Kosten des Straßengüterverkehrs. Stellungnahme des Wissenschaftlichen Beirats beim Bundesminister für Verkehr, Bau und Stadtentwicklung. *Zeitschrift für Verkehrswissenschaft, 81*(2), 73–105.

Ahrens, G.-A., & Hubrich, S. (2012). Mobilitätsmanagement—zentrales Element einer integrierten Verkehrsentwicklungsplanung. In M. Stiewe & U. Reutter (Eds.), *Mobilitätsmanagement. Wissenschaftliche Grundlagen und Wirkungen in der Praxis* (pp. 30–48). Essen: Klartext.

Allen, J., Browne, M., Piotrowska, M., & Woodburn, A. (2010). *Freight quality partnerships in the UK—An analysis of their work and achievements*. London: Transport Studies Group, University of Westminster.

Allen, J., Thorne, G., & Browne, M. (2008). *BESTUFS. Good practice guide on urban freight transport*. Rijswijk/Karlsruhe: NEA Transport Research and Training/PTV, Planung Transport Verkehr.

Anderson, S., Allen, J., & Browne, M. (2005). Urban logistics—How can it meet policy makers' sustainability objectives? *Journal of Transport Geography, 13*(1), 71–81.

Ballantyne, E. E. F., Lindholm, M., & Whiteing, A. (2013). A comparative study of urban freight transport planning: Addressing stakeholder needs. *Journal of Transport Geography, 32*(10), 93–101.

Binnenbruck, H. H., Hoffmann, P., & Krug, S. (1998). Mobilitätsmanagement im Personen- und Güterverkehr. *Straßenverkehrstechnik, 42*(9), 19–24.

Blees, V. (2010). Mobilitätsmanagement und Städte. *Der Nahverkehr, 28*(1-2), 61–65.

BMVBS – Bundesministerium für Verkehr, Bau und Stadtentwicklung. (2010). *Aktionsplan Güterverkehr und Logistik—Logistikinitiative für Deutschland*. Berlin.

Boltze, M. (1996). Intermodales Verkehrsmanagement. *Internationales Verkehrswesen, 48*(1+2), 11–18.

Boltze, M. (2013). Transportmanagement: Güterverkehrsnachfrage stadtverträglich beeinflussen. In Deutsches Institut für Urbanistik (Ed.), *Urbane Räume in Bewegung. Geschichte, Situation und Perspektive von Stadt* (pp. 277–286). Berlin: Deutsches Institut für Urbanistik.

Boltze, M., Friedrich, H., & Rühl, F. (2012, March 14). *Freight transport demand management: A contribution to urban traffic management and sustainability of logistics*. WCTRS SIG10 Workshop. Vienna.

Boltze, M., Jiang, W., Groer, S., & Scheuvens, D. (2014). Analyse der Wirksamkeit von Umweltzonen hinsichtlich Feinstaub- und Stickstoffkonzentrationen. *Straßenverkehrstechnik, 58*(4), 219–228.

Börjesson, M., Eliasson, J., Hugosson, M., & Brundell-Freij, K. (2012). The Stockholm congestion charges—5 years on. Effects, acceptability and lessons learnt. *Transport Policy, 20*(3), 1–12.

Bretzke, W.-R., & Barkawi, K. (2012). *Nachhaltige Logistik*. Berlin: Springer.

Browne, M., Allen, J., Nemoto, T., Patier, D., & Visser, J. (2012). Reducing social and environmental impacts of urban freight transport: A review of some major cities. *Procedia—Social and Behavioral Sciences, 39*, 19–33.

Bruns, A., & Langweg, A. (2010). Mobilitätsmanagement—Konzept für die Umsetzung in der kommunalen Praxis. In T. Bracher, K. Dziekan, J. Gies, H. Holzapfel, F. Huber, F. Kiepe, et al. (Hg.), *HKV—Handbuch der kommunalen Verkehrsplanung. Strategien, Konzepte, Maßnahmen für eine integrierte und nachhaltige Mobilität* (Chapter 3.1.3.4). Berlin/Bonn: Wichmann/Economica-Verlag anfangs.

BVU Beratergruppe Verkehr + Umwelt GmbH, Intraplan Consult, Ingenieurgruppe IVV, & Planco Consulting. (2014). *Verkehrsverflechtungsprognose 2030. Los 3: Erstellung der Prognose der deutschlandweiten Verkehrsverflechtungen unter Berücksichtigung des Luftverkehrs.*

Castro, J. T., & Delos Reyes, M. R. (2010). Estimating traffic and emissions for various scenarios of freight vehicle restrictions in metro manila. *Asian Transport Studies, 1*(1), 4–17.

Dablanc, L. (2007). Goods transport in large European cities: Difficult to organize, difficult to modernize. *Transportation Research Part A: Policy and Practice, 41*(3), 280–285.

Dablanc, L. (2011). *City logistics best practices: A handbook for authorities* (SUGAR sustainable urban goods logistics achieved by regional and local policies). Bologna: Interreg IVC.

Dablanc, L., Guiliano, G., Holliday, K., & O'Brien, T. (2014). Best practices in urban freight management. Lessons from an international survey. *Transportation Research Record: Journal of the Transportation Research Board, 2379*, 29–38.

Dezi, G., Dondi, G., & Sangiorgi, C. (2010). Urban freight transport in Bologna: Planning commercial vehicle loading/unloading zones. *Procedia—Social and Behavioral Sciences, 2* (3), 5990–6001.

EPOMM - European Platform on Mobility Management. (2013). *Mobility management. The smart way to sustainable mobility in European countries, regions and cities.* Brüssel.

FGSV - Forschungsgesellschaft für Straßen- und Verkehrswesen. (2012). *Begriffsbestimmungen—Teil: Verkehrsplanung, Straßenentwurf und Straßenbetrieb.* Köln: FGSV-Verlag.

Fiedler, J. (2002). Mobilitätsmanagement als Chance. *Der Nahverkehr, 20*(1–2), 23–26.

Fischer, M. J., Ahanotu, D. N., & Waliszewski, J. M. (2003). Planning truck-only lanes: Emerging lessons from the Southern California experience. *Transportation Research Record: Journal of the Transportation Research Board, 1833*, 73–78.

Gertz, C., Altenburg, S., & Bohnet, M. (2012). Ex-ante-Quantifizierung der Wirkung von Mobilitätsmanagementmaßnahmen. In M. Stiewe & U. Reutter (Eds.), *Mobilitätsmanagement. Wissenschaftliche Grundlagen und Wirkungen in der Praxis* (pp. 198–213). Essen: Klartext.

Givoni, M. (2012). Re-assessing the results of the london congestion charging scheme. *Urban Studies, 49*(5), 1089–1105.

Government of the Netherlands. (2015). Green deal. Online verfügbar unter. zuletzt geprüft am 02.07.2015, from http://www.government.nl/issues/energy-policy/green-deal

Holguín-Veras, J., Ozbay, K., Kornhauser, A., Brom, M. A., Iyer, S., Yushimito, W. F., et al. (2011). Overall impacts of off-hour delivery programs in New York City Metropolitan area. *Transportation Research Record: Journal of the Transportation Research Board, 2238*, 68–76.

ILS/ISB – Institut für Landes- und Stadtentwicklungsforschung des Landes Nordrhein-Westfalen; Institut für Stadtbauwesen. (2000). *Mobilitätsmanagement Handbuch.* Dortmund/Aachen.

Kemming, H. (2009). Mobilitätsmanagement—Perspektivenwechsel in der Verkehrspolitik. In Bundesministerium für Verkehr, Bau und Stadtentwicklung (Ed.), *Urbane Mobilität. Verkehrsforschung des Bundes für die kommunale Praxis* (pp. 377–396). Bremerhaven: NW Verl. für neue Wissenschaft.

Kittler, W., & Boltze, M. (2011). Beeinflussung der Zeitwahl von Nutzern des ÖPNV. *Der Nahverkehr, 29*(6), 7–12.

Köhler, U. (1999). City logistics in Kassel. In E. Taniguchi & R. G. Thompson (Eds.), *City logistics I* (pp. 261–271). Kyoto: Institute for City Logistics.

Lindholm, M., & Blinge, M. (2014). Assessing knowledge and awareness of the sustainable urban freight transport among Swedish local authority policy planners. *Transport Policy, 32*(3), 124–131.

Lindholm, M., & Browne, M. (2013). Local authority cooperation with urban freight stakeholders: A comparison of partnership approaches. *European Journal of Transport and Infrastructure, 13*(1), 20–38.

Litman, T. (2013). Transportation demand management: Win-Win solutions to transport problems. In G. Wulfhorst (Ed.), *Transportation demand management. Insights from the mobil. TUM 2012, International Scientific Conference on Mobility and Transport* (pp. 3–9). München: Inst. für Verkehrswesen.

Lucietti, L. (2012). *C-LIEGE toolbox for the establishment of the city logistics manager*. C-LIEGE Deliverable n. 4.2.

Manheim, M. L. (1979). *Fundamentals of transportation systems analysis*. Cambridge, MA: MIT Press.

Menges, K., & Boltze, M. (2016). Temporal shifts in freight traffic—Estimating the potential to improve traffic quality on German highways with temporal shifts of heavy traffic. In E. Abele, M. Boltze, & H.-C. Pfohl (Eds.), *Dynamic and seamless integration of production, logistics and traffic. Fundamentals of interdisciplinary decision support*. Berlin: Springer.

Miltner, T. (2011). Lkw-Empfehlungsnetz Region Frankfurt RheinMain. In U. Clausen (Ed.), *Wirtschaftsverkehr 2011: Modelle—Strategien—Nachhaltigkeit* (pp. 113–125). Dortmund: Verlag Praxiswissen.

Muñuzuri, J., Larrañeta, J., Onieva, L., & Cortés, P. (2005). Solutions applicable by local administrations for urban logistics improvement. *Cities, 22*(1), 15–28.

Ogden, K. W. (1992). *Urban goods movement. A guide to policy and planning*. Aldershot: Ashgate.

Östlund, B., Janson, G., Burgess, A., Ruijgrok, K., Tavasszy, L., & Steen Petersen, M. (2002). *A logistics module for SAMGODS*. Stockholm: Swedish Institute for Transport and Communications Analysis.

Quak, H. J. (2008). *Sustainability of urban freight transport. Retail distribution and local regulations in cities*. Delft: Netherlands TRAIL Research School.

Quak, H. J., & de Koster, M. B. M. (2009). Delivering goods in urban areas: How to deal with urban policy restrictions and the environment. *Transportation Science, 43*(2), 211–227.

Reutter, U. (2014). Mobilitätsmanagement: ein Beitrag zur Gestaltung einer nachhaltigen Mobilität. In T. Bracher, K. Dziekan, J. Gies, H. Holzapfel, F. Huber, F. Kiepe, et al. (Hg.), *HKV—Handbuch der kommunalen Verkehrsplanung. Strategien, Konzepte, Maßnahmen für eine integrierte und nachhaltige Mobilität* (Chapter 1.2). Berlin/Bonn: Wichmann/Economica-Verlag anfangs.

Rühl, F., Mörner, M. V., Friedrich, H., & Özsucu, Ö. (2013). Impacts of HGV tolls on transport logistics. In *General Proceedings of the 13th World Conference on Transport Research Society*.

Rühl, F., & Ottemöller, O. (2013). Nachfragebeeinflussung im Güterverkehr. In Institut für Stadtbauwesen und Stadtverkehr, RWTH Aachen (Eds.), *Tagungsband der "AMUS + ACMoVe 2013 User-friendly Mobility—Trends, Challenges and Innovations"* (pp 35–37).

Russo, F., & Comi, A. (2011). Measures for sustainable freight transportation at urban scale: Expected goals and tested results in Europe. *Journal of Urban Planning and Development, 137* (2), 142–152.

Son, N. N. (2006). Transportation in and around Hanoi: Current issues and challenges. In K. Ohno & V.-T. Nguyen (Eds.), *Business environment and policies of Hanoi* (pp. 63–121). Tokyo: Publishing House of Social Labour.

SRU – Sachverständigenrat für Umweltfragen. (2005). *Umwelt und Straßenverkehr. Hohe Mobilität—Umweltverträglicher Verkehr*. Köln: Bundesanzeiger Verlagsgesellschaft.

SRU – Sachverständigenrat für Umweltfragen. (2012). *Umweltgutachten 2012. Verantwortung in einer begrenzten Welt*. Berlin: Erich Schmidt Verlag.

Stathopoulos, A., Valeri, E., & Marcucci, E. (2012). Stakeholder reactions to urban freight policy innovation. *Journal of Transport Geography, 22*(5), 34–45.

Taniguchi, E., James, J., Barber, R., Imanishi, Y., & Debauche, W. (2012). *Public sector governance of urban freight transport*. PIARC World Road Association.

Taniguchi, E., & Nemoto, T. (2002). Transport-demand management for freight transport. In E. Taniguchi & R. G. Thompson (Eds.), *Innovations in freight transport* (pp. 101–124). Southampton: WIT.

Tavasszy, L., & Ruijgrok, K. (2013). Freight transport: Indicators, determinants and drivers of change. In G. P. van Wee, J. A. Annema, & D. Banister (Eds.), *The transport system and transport policy. An introduction* (pp. 51–77). Cheltenham: Edward Elgar.

Torentellé, M., Tsamboulas, D., & Moraiti, P. (2012). *Elicitation of the good practices on urban freight transport*. C-LIEGE Deliverable n. 2.1.

UBA – Umweltbundesamt. (2001). *Mobilitätsmanagement zur Bewältigung kommunaler Verkehrsprobleme*. Berlin.

UBA – Umweltbundesamt. (2009). *Strategie für einen nachhaltigen Verkehr*. Dessau.

Van Rooijen, T., & Quak, H. (2014). City logistics in the European CIVITAS initiative. *Procedia—Social and Behavioral Sciences, 125*, 312–325.

VTPI – Victoria Transport Policy Institute. (2015). *Freight transport management. Online TDM Encyclopedia*. Accessed July 29, 2015, from www.vtpi.org

Wiegel, F., Adolph, S., Özsucu, Ö., Thiel, D., Abele, E., & Elbert, R. (2013). Urbane Wertschöpfung. Herausforderungen und Potenziale für Produktion und Logistik im urbanen Umfeld. *Industrie Management, 29*(5), 15–18.

Implications for Freight Transport Demand Modelling from Interdisciplinary Research: Developing a Concept for Modelling Freight Transport Within Supply Networks of the Automotive Industry

Ole Ottemöller and Hanno Friedrich

Introduction

The transport system plays a key role for national and international economic prosperity by offering mobility to people and goods. Hence, it bridges the spatial gap between locations of production and consumption for raw materials, intermediates and final products. With today's multi-tier and minimum-inventory supply chains, the transport system's reliability becomes a competitive factor for national economies. Together with the circumstance of limited budgets and strict environmental regulations, public authorities face limited room for manoeuvre for creating and maintaining a sustainable transport system that fulfils the manifold arising requirements. In order to cope with these challenges, reliable prognoses and policy assessments are needed. These are the key application areas of freight transport models. Thus, when developing these models two questions arise: What are the most important drivers of freight transport? How do we capture these drivers in freight transport models? In the following, we are going to structure the solution space of these questions and give answers for selected aspects.

All of the eligible drivers can be aggregated to so-called megatrends. Over the last decades, these megatrends have strongly influenced the development of society and economy. Not only certain industries have vanished or moved to other countries (e.g. German coal mining) but also new industries have emerged (e.g. mobile

O. Ottemöller (✉)
Institute of Traffic and Transport, Technische Universität Darmstadt, Otto-Berndt-Str. 2, 64287 Darmstadt, Germany
e-mail: ottemoeller@verkehr.tu-darmstadt.de

H. Friedrich
Wissenschaftliche Hochschule für Logistik und Unternehmensführung, Kühne Logistics University – KLU, Großer Grasbrook 17, 20457 Hamburg, Germany
e-mail: hanno.friedrich@the-klu.org

E. Abele et al. (eds.), *Dynamic and Seamless Integration of Production, Logistics and Traffic*, DOI 10.1007/978-3-319-41097-5_10

telecommunication). Other industries have adapted their processes and products to the arising new challenges and opportunities. The automotive industry has been one of these highly dynamic industries as it has successfully implemented new concepts in production and logistics. Furthermore, it has also integrated new technologies into the produced vehicles. Hence, we will use the German automotive industry as an example for the hypotheses discussed.

In the first section, we deal with rather general challenges that accompany modelling in the interdisciplinary domain of production, logistics and transport. Next, we move on to the phenomenon freight transport, suitable indicators for its description and drivers influencing freight transport demand. In the following section, we describe the German automotive industry from our perspective as freight transport demand researchers point out potential future developments for the industry. As most developments are correlated with effects influencing the supply chain structure, we focus on the interrelation of supply chain structure and freight transport demand in the subsequent section. Afterwards, we introduce data sources that can be used for analysis in this context and discuss first evidence for the existence of the effects described. Based on this insight, we present a conceptual model for the procedural generation of supply chains on the basis of business establishments that can be integrated into freight transport demand models. Finally, we close with a brief conclusion and outlook.

Modelling in the Domain of Production, Logistics and Transport

In general, models are simplified representations of reality (Simon 1996). These models can be of descriptive or explanatory nature. Explanatory models can be categorised as either normative or positive regarding their relation to real-world systems (Bossel 2004). In our interdisciplinary context, models can further be distinguished by referring to the following three aspects:

- Objective (Why?)
- Scope (What?)
- Method (How?)

Before applying or developing a model, a clear definition of the model's objective is needed. If we do not know the question, it is impossible to come up with a suitable answer. Once the objective has been set, we can define the borders of our analysis. In other words, we limit the scope of our model in accordance with the intended objective. The definition of the model scope usually incorporates information on space and time as well as on objects and activities. The latter is closely related to the discussion on which decisions to include in the model as these determine the internal degree of freedom. Besides that, additional limitations, e.g. regarding scale, resolution and accuracy, may contribute to the scope definition.

Once we have defined why and what to model, we can continue by deciding on how to model. Hence, in this step we determine which methods we are going to use in the intended model. Complex models, such as freight transport demand models, might combine descriptive as well as explanatory methods. For the actual model creation, once objective, scope and methods are set, we recommend to follow the instructions given in more specific literature on that topic (see for instance Page and Kreutzer 2005 or Law 2006).

Before taking a look at the interdependencies of production, logistics and freight transport, we need a more precise definition of these terms. Therefore, we refer to the value chain concept introduced by Porter (1985). In our context, production subsumes all aspects related to the design of processes within a manufacturing firm's operations. This ranges from decisions regarding the organisation of internal material flows to decisions on working shifts and machine setups on the shop floor. In Porter's value chain, inbound and outbound logistics are responsible for the management of material and information flows across plant borders. As both activities cope with linking business establishments to business establishments or consumer markets, we simplify and only use the aggregate term logistics for all planning activities concerning these flows. From a larger perspective, the entirety of all value chains constitutes freight transport. Again, freight transport itself is part of the overall transport system, which also includes passenger transport.

Looking at the defined research disciplines of production, logistics and transport, we see major differences between the models applied. Accompanying the obvious differentiating factors in objective and method, we find a very strong separation in the scope, especially the model scale. Freight transport demand models aim at describing or predicting the number of vehicles on the infrastructure caused by any freight movement. The models' spatial scopes range from urban to national or international contexts. Due to their broad scope, we define these models to be working on a macro level. In contrast, models from production, i.e. process simulations, often focus on single assembly lines within plant boundaries integrating very detailed information. Hence, they work on a micro level. Between these two, we find logistics models, which traditionally contain multiple network locations but only those that are in the model user's planning scope. This view represents a meso level. Still, reality behind these models is not as disjunct and independent as these models suggest. For example outsourcing of production processes, resulting from the identification of inefficiencies on the plant level, leads to additional demand in logistics and freight transport. The opposite direction of action is found if higher volatility in delivery times, caused by effects in the transport system, leads to adaptations in the inbound inventories of production. Obviously, certain applications require models that are capable of dealing with such interdisciplinary influences. It is important to keep in mind that the translation from micro to macro level cannot be achieved by simple additive or multiplicative aggregation as we would face two problems: the first being unsolvable complexity when upscaling isolated microscopic process models. The second and even more severe problem is the so-called micro–macro gap in freight transport modelling caused by vehicle movements and commodity flows not being identical due to consolidation efforts in logistic networks (Liedtke and Friedrich 2012).

Nevertheless, it is necessary to integrate the impact of effects coming from the other subsystems in the different models. Our work especially aims at integrating these aspects into freight transport demand modelling. Actually, this idea is not totally new to transport research. The high interdependency of the overall transport system, societal as well as economic activity and resulting flows was first introduced by Manheim (1980). Unquestionably, this still holds true especially for freight transport as it is a result of manufacturing and logistics activity.

Freight Transport and its Drivers

Freight transport does not create utility by itself (Ortúzar and Willumsen 2011). It is a derived demand that results from the spatial gap between production and consumption of goods. Therefore, freight transport analysis must incorporate the underlying activity systems' behaviour. Also, freight transport is an aggregate phenomenon. It is constituted by the movement of vehicles along infrastructure for the transport of goods. Regulation and infrastructure supply are often centrally organised by governmental authorities for this system. In contrast, the operation of vehicles under given conditions is planned and carried out by decentralised units. Among the manifold degrees of freedom, the most important ones might be the decision on the transport volume, the spatial relation (i.e. origin and destination of shipments) and the transport chain choice (i.e. enhanced mode choice). As freight transport researchers, we are interested in aggregated key indicators for analyses regarding the overall freight transport system's performance. These indicators are calculated on the basis of surveys on the goods transported and resulting vehicle movements. The following list of selected indicators is partly taken from McKinnon (2008):

- Freight transport volume (t)
- Freight transport performance (tkm)
- Vehicle performance (laden/empty running) (vkm)
- Vehicle capacity utilisation (%)
- Average length of haul (km)
- Modal split (%)

Following a top-down approach, these indicators can be further disaggregated, e.g. by vehicle categories or types of goods. For Germany, this data is published in an annual series (Bundesministerium für Verkehr, Bau und Stadtentwicklung 2013). Such aggregate data can already be used for first analyses of freight transport developments as shown in Tavasszy and Ruijgrok (2013).

In order to do temporal and spatial analyses of freight transport leading far beyond available aggregate data, more complex freight transport models are required. First transport demand models were developed for the application in passenger transport (Martin et al. 1965; Hutchinson 1974). Subsequently, based on the framework of passenger transport, freight transport models have been introduced, successively including specific aspects like trade or logistics. Models addressing logistics, especially at the urban level, represent the latest manifestation of this development. Good overviews of the existing freight transport demand models can be found in Chow et al. (2010), Tavasszy et al. (2012) and de Jong et al. (2013) for the national level or in Taniguchi et al. (2014) for the urban level.

As it interacts with dynamically changing activity systems, freight transport itself evolves as well. In the past, slowly developing trends, e.g. the shift from production-based to service-based economies, as well as radical changes, e.g. the fall of the iron curtain, have influenced the shape of freight flows. Currently, we can observe the maturation of ongoing trends as well as the emergence of new trends. We can classify the trends influencing freight transport by their origin: trade and production, the transport system or regulation.

Trade and production are responsible to fulfil customer demand, caused by private as well as company consumption of materials. Here, we expect changes related to production technology leading to new spatial setups of supply networks. For example, affordable small- to mid-scale automation solutions enable the establishment of smaller production facilities in close proximity to demand locations instead of running single centralised production plants, thereby reshaping the spatial flow of intermediates and final products. The spatial layout of multistage supply chains is also influenced by ongoing processes of vertical disintegration due to specialisation of suppliers accompanied by internationalisation, often triggered by wage differences. Besides that, product innovations have the power to change the flow of goods, as they will require different raw materials and intermediates from different regions than before. In this regard, one example is the production of fully electric cars that do contain an electric powertrain, which gets assembled at a different location than the combustion-based solution used before. As a result, new trade patterns emerge. Independently, private customers change their shopping behaviour leading to new commodity distribution structures.

The transport system itself does also face new developments. Coming from recent advances in information technology, real-time analytics enabled by big data will allow for better traffic control and better information for transport system users, e.g. freight carriers. Also, there are plans, especially in developing countries, for large investments in new infrastructure. These new connections in the existing transport system will foster the accessibility of remote regions and therefore directly impact national as well as international freight flows. Also, the broad application of new technology, e.g. automated guided vehicles, might lead to radical changes in the transport system in the way we can already today see the impact of navigation systems on infrastructure usage.

Last but not least, regulation will have an increasingly strong influence on freight transport. After focusing on passenger mobility, public authorities start putting effort into freight transport demand management in order to mitigate requirements from businesses and traffic-oriented targets. This trend is intensified by the need for measures that help to achieve sustainability targets, especially related to environmental protection, as existing agreements demand for major reductions in greenhouse gas emissions. Another focus of governmental action will need to be on the security of supply chains and risk prevention for the population as international supply chains are increasingly vulnerable to disruptions caused by political instabilities or extreme events, e.g. caused by nature, epidemics or technical failure.

The impact of these trends on freight transport heavily depends on the industry under consideration. Hence, we recommend using freight transport models that focus on a specific economic sector, industry. The idea of limiting a freight transport model's scope in this way has its origin in the main challenges of transport modelling, i.e. data availability, homogenous groups of actors and decisions and complexity. Accordingly, we work on creating models that capture the most important drivers of freight transport demand in the industry considered. We propose to refer to the following set of assertions to define sectoral models:

Sectoral models limit their scope to one economic sector. Their objective is the analysis of the sector‘s system behaviour. Hence, they focus on a single sector‘s characteristics, i.e. system elements, interrelations and especially actors and their decisions.

In the following, we elaborate on aspects relevant for the application of sectoral freight demand modelling to the German automotive industry.

The German Automotive Industry from the Perspective of Freight Transport Demand Modelling

Analysing the regional distribution of automotive activity in Germany, several local clusters can be identified, which strongly depend on the prosperity of this sector.

The intensively applied concept of distributed value creation leads to a high demand for transport. Modern provisioning concepts like just-in-time (JIT) and just-in-sequence (JIS) deliveries lead to a strong dependence on the transport system's reliability. Therefore, transport infrastructure and services are competitive factors for this industry, especially in the global competition. So far, research has been conducted on different levels of aggregation like individual commodity flows, supply chains or total freight transport demand. A study on how the spatial setup influences the performance of automotive supply networks has been conducted by Dyer (1996). Currently available statistics and freight transport models do not allow for analysis of changes in freight transport. A sound representation on the sector level as an aggregation of the vast number of single supply chains still is missing.

On the national level, the automotive industry accounts for 5–15 % of the total employment (Tillmann 2009). This wide range results from the difficulty of defining borders for a certain industry and the underlying objective from interest groups when estimating these numbers. Our approach to define automotive industry starts from the car plants and then stepwise follows inbound flows upstream till a level of

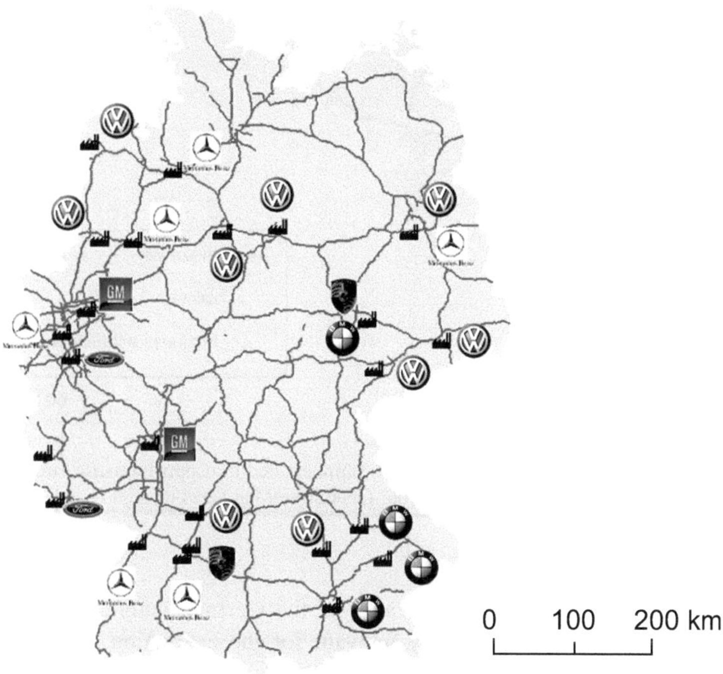

Fig. 1 Map of Germany with car plant locations by OEM together with motorway network. Data from OpenStreetMap and Automobil Produktion (2008)

generic supplies is reached. This definition is wider than those used in current statistics.

The same problem of clearly identifying the automotive industry in statistics holds true for freight transport. In the national transport statistics, the category *vehicles and parts thereof* accounts for a share of 4 % as part of the total annual freight transport performance on road as well as rail measured in tonne-kilometres (Statistisches Bundesamt 2013a, b). The map in Fig. 1 shows an overlay of the German motorways and the location of car plants. Here, a strong adjacency between production facilities and transport infrastructure can be observed. This fact supports the assumption of correlation between transport and automotive production. Obviously, the emergence of the overall structure underlies many more influences.

Regarding the locations of all brands' car plants in Germany, we cannot identify a single cluster of car production. Instead, regional agglomerations of factories are spread over the country. Naturally, the regions around car plants strongly depend on the employment effects generated by the production of cars. Going into detail for a single region, Southern Hesse in this case, the mutual affinity of transport infrastructure and business establishments of the automotive industry can also be observed on the more detailed level. This correlation is visualised in Fig. 2.

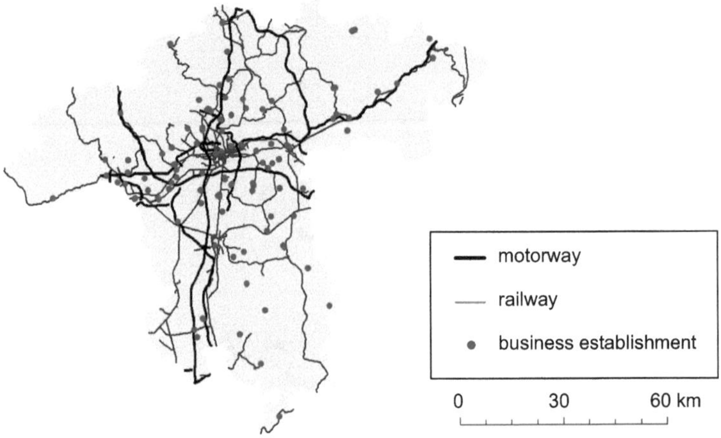

Fig. 2 Business establishments of automotive suppliers and transport infrastructure in Southern Hesse, Germany. Data on infrastructure from OpenStreetMap and business establishments from proprietary data

In the following paragraphs, we analyse the German automotive industry in regard to characteristics potentially relevant for the estimation of future freight transport demand.

After claiming the patent for the construction of the first automobile with a combustion engine in 1886, the name Benz is still well known in the car market nowadays. In the meantime, manifold new companies and brands have appeared in and also vanished from the automotive industry. For Germany, there are a number of multicorporate enterprises producing cars in Germany, some of them having multiple brands in their portfolio. The largest of these is the Volkswagen Group, combining brands like VW, Audi, Porsche and many more under one roof. In total, there are approximately 30 car plants in Germany run by the different OEM. Although collaboration exists in some areas, the car producing enterprises rely on distinctly organised supply and transport networks. Hence, stakeholders can be independent or assigned to one or more brand networks, leading to different planning scopes.

The final car assembly in the actual car plant is the last step of a long chain of distributed value creation. Typically, several key processes are located on the car plant's compound as they require enormous process knowledge and transport of the intermediate output to the assembly is difficult. In most cases, this is the press shop, body shop and paint shop (Ihme 2006). Still, there are exceptions to the rule, e.g. in environments that make use of completely knocked-down concepts (CKD).

Parts and components needed for the final assembly are produced in a pyramid-shaped supply network with the car plant being located at the top. On the first level so-called first-tier suppliers assemble components to modules, which are delivered directly to the assembly line where they are put into the vehicle in a job flow system.

The number of modules per vehicle is limited and almost identical between most vehicles if seen from our level of abstraction.

As most cars are customer specific and therefore build to order, it is a demanding but also crucial task to supply the needed parts to the right station of the assembly line in the correct order and at the needed time. As a result, the automotive industry developed advanced logistics systems for internal and external material flows. In the case of first-tier suppliers, this leads to a large number of modules being delivered by truck as direct JIT or JIS transports to the assembly line. The transport strategy as well as the tight collaboration with the OEM influences the location choice of suppliers. Thus, many first-tier suppliers run business establishments near the car plant to minimise the risk of delays due to traffic disturbances. One measure against transport vulnerabilities is joining so-called supplier parks. These are located in direct neighbourhood to the car plant and offer space to supplying firms for their plant-specific production (Becker 2006).

According to the hierarchy-based nomenclature, the described first-tier suppliers get their inputs from second-tier suppliers, which produce automotive-specific products from more generic inputs. On this level, the distinct assignment of business establishments and their outputs to single car plants or even brands is nearly impossible.

Obviously, the decision where to locate the production of supplying parts is mainly driven by minimisation of total costs. Therefore, labour-intensive production gets shifted to low-cost countries, increasing transport costs and transport vulnerability, especially in JIS/JIT environments. For example, several suppliers moved their assembly of cable harnesses to Tunisia still offering JIS/JIT deliveries (Holweg and Pil 2004). Not only northern Africa experiences a growth in the number of vehicle component suppliers but also Eastern Europe benefits from the OEM's rising cost sensitivity (VDA 2004). However, the German input–output table shows that still most intermediate products that are consumed as inputs by the automotive industry come from national production if measured by value. Expert interviews indicated that most imported goods are characterised as high technology or labour-intensive products, which usually strongly differ in their physical attributes. Nevertheless, we can derive that imports play an increasingly important role in OEM sourcing. Depending on the good considered these markets can be local, national, regional (regional standing for regions on a global level like Europe or Northern America) or global.

The further distribution of value creation also leads to innovative transport solutions. For example a major share of Audi's engines is produced in Győr (Hungary) and built into the vehicles at different plants in the widespread Volkswagen production network (Audi 2013). Most of the transports leaving the engine factory in Győr rely on block trains delivering the built engines directly to the Audi plant in Ingolstadt (Germany) (Audi 2008). Following this trend, several OEM have declared the target of shifting more transports from road to rail in the future combining economic and environmental benefits.

In summary, we have pointed out certain characteristics of the German automotive industry and briefly shown the connection to the resulting freight transport demand.

Potential Developments for the German Automotive Industry

Now, we are going to take a look at potential drivers influencing the future development of freight transport demand in the automotive industry. Whenever the complexity of the underlying system and its environment does not allow for a closed holistic model, using scenarios can help to capture uncertainties (Schoemaker 1993). In our context, these scenarios can contain one or more assumptions on the future development of motor vehicle production, which is tightly connected to the resulting freight transport demand. A selection of potential trends based on experts' opinion and our own considerations is described below.

Strict environmental regulations and customers' demand for increased energy efficiency will force car manufacturers to build vehicles with less output of CO_2 and particulate matter (e.g. European Commission 2011). Already today the number of new registrations for electric and hybrid vehicles is increasing strongly, although on a very low level (KBA 2013). However, in the next decades conventional combustion engines are keen to be successively replaced by alternative powertrains. This change will also have a strong impact on production networks and the transports within since the production process of current powertrain technology is substantially different from its electric alternative. With the target of reduced fuel consumption, car producers develop technologies regarding lightweight constructions like carbon bodies. Thus, the flow of intermediates into the automotive sector as described in the national accounting's input–output matrix will shift, e.g. from metal-bound industries towards industries relying on lightweight materials like carbon fibre.

Customers' higher cost sensitivity and changes in mobility behaviour might lead to smaller and more standardised vehicles. Hence, cars would become mid-price commodities losing complexity and specificity in the product and the production process potentially allowing for new transport optimisations (Katayama and Bennett 1999).

Urbanisation and changing mobility needs might also lead to a strong decrease in demand for automobiles on a European or even global level. Significant drops in demand would force car producers to fundamentally redesign their production networks and totally change the landscape of vehicle producing firms and their supply chains.

For the general structure of the industry, multiple changes are expected. For the so-called Original Equipment Manufacturers (OEM) as well as suppliers a trend towards stronger market concentration can already be identified as competitors are forced out of the market due to cost pressure and takeovers (Holweg and Pil 2004). New brands from Asia bring some new heterogeneity, but at the same time their entry into global competition strengthens the price pressure as Europe already has excess capacity in the production of automobiles (Güttner and Sommer-Dittrich 2008; PwC 2012).

The future location and number of car plants may differ from today's layout due to shifts in costs or other uncertainties like availability and reliability of infrastructure (MacCarthy and Atthirawong 2003). Under conditions of rising freight transport in the road system and higher road transport costs, modal shifts towards rail or waterway for inbound transport might become desirable for the automotive industry. In order to achieve sustainability targets, governments could even force companies to shift their transports away from road by regulatory measures.

New developments in production technology might also lead to major changes. A lot of research and development is currently invested into modular production systems that allow for smaller lot sizes, better flexibility and transportability of production facilities (Rogers and Bottaci 1997). Our discussions with experts have shown that for selected parts this trend would lead to multiple small manufacturing locations in direct neighbourhood to the car plant instead of large single factories supplying multiple OEM.

Most of the potential developments described above lead to changes in the automotive supply chains. Hence, in the following section we will focus on the relation of supply chain structures and freight transport demand.

Supply Chain Structure Effects as Drivers of Freight Transport Demand

Today, it is broadly accepted that the competitive advantage depends on the excellence of all activities within a firm. Porter (1985) describes this principle and introduces the value chain concept, which structures material-flow-based firms' processes into strategic activities. Building upon these ideas, Christopher (1992) further develops this idea by clarifying that it is no longer isolated firms competing against each other. Instead, he puts networks of interdependent companies, i.e. supply chains, in this role. Hence, in the last years, companies have spent much effort on the optimisation of supply chains. Often, this leads to vertical or horizontal restructuring. Vertical effects address actors of consecutive stages of the supply chain, e.g. outsourcing of processes, while horizontal effects take place between actors of the same supply chain stage, e.g. takeovers or alliances (Werner 2013). Derived from their nature, vertical effects can be described as (dis)integrations while (dis)aggregations correspond to horizontal effects. Some manifestations of such restructurings are of organisational nature, having no direct impact on physical flows. As we are interested in freight transport, in the following we will focus on supply chain effects that manifest on the physical level. Here, physical level stands for material flows between business establishments, which are situated at defined locations. The idea of including these aspects is not totally new to the field of freight transport demand modelling. The ADA model by de Jong and Ben-Akiva (2007) was the first fully applied freight transport model that explicitly considers relations between shipping and receiving firms. A general framework to describe interrelations on this level regarding freight transport issues has been presented by Roorda et al. (2010). Their actor-based framework specifies a standard of how organisational and physical units interact focusing on business establishments and their configurations. A first effort to integrate detailed supply chains into freight transport models has been undertaken in the FAME model by Samimi et al. (2010). Our approach is consistent with the before-mentioned works but differs in the objective as we are more interested in the description and impact analysis of certain sectoral supply chain effects on the level of flows between business establishments.

An often-cited indicator for the ongoing change within industries and national economies is the value added (OECD 2013) or the ratio of value added per output in monetary units (calculated as ratio $= \frac{\text{output}-\text{input}}{\text{output}}$). Looking at the German automotive industry, we see a significant decline of this indicator between 1967 and 2000, stabilising around 20 % in the following years as depicted in Fig. 3. The potentials and possible limitations for future development are still under discussion (Göpfert and Grünert 2012).

As mentioned before, we are interested in supply chain effects of horizontal and vertical nature. If we did remain with these two categories, we did not cover relocations of business establishment that are part of supply chains. Hence, we have to extend our categories and also include spatial effects. A real-life example for this type of effect is outsourcing in case previous in-house production gets shifted across national borders. Spatial effects can also take place within national borders, e.g. if production plants follow governmental subsidies or customers.

As the effect of decreasing ratio of value added per output and its potentially negative impact on freight transport has been in the public debate quite some time without scientifically satisfying answers (e.g. Holzapfel and Vahrenkamp 1999), we

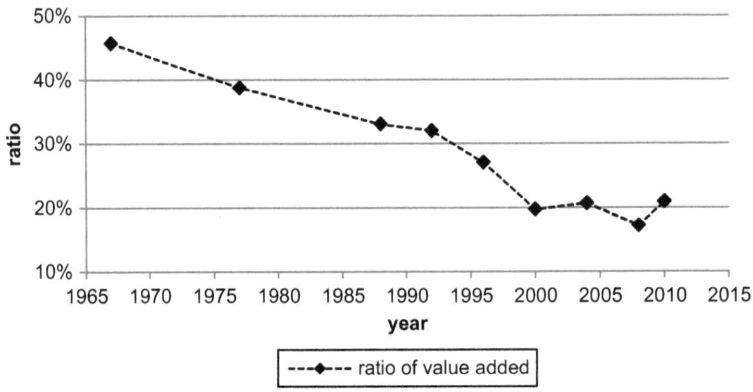

Fig. 3 Development of value added per gross production in the automotive industry. Data taken from VDA (1969–2012)

try to shed some light by introducing small schematic numerical scenarios for possible supply chain effects and their impact on freight transport and the related value added per output. The scenarios and key values are illustrated in Fig. 4.

Our base scenario (a) consists of three business establishments connected by a simple supply network, in this first case a linear supply chain. The chain is driven by the demand of a consuming unit, e.g. a vehicle manufacturer's production site, which is supplied via a first-tier supplier that gets its inputs from a second-tier supplier. The latter procures required raw materials from abroad. The dotted line denotes an international border as our system boundary. All values correspond to monetary units. Obviously, this is only of limited usefulness in regard to freight transport, but it helps keeping the examples simple and connected to available statistics. How to convert these monetary units into material flows and later vehicle flows will be discussed in the data. Getting back to our base scenario, we can now

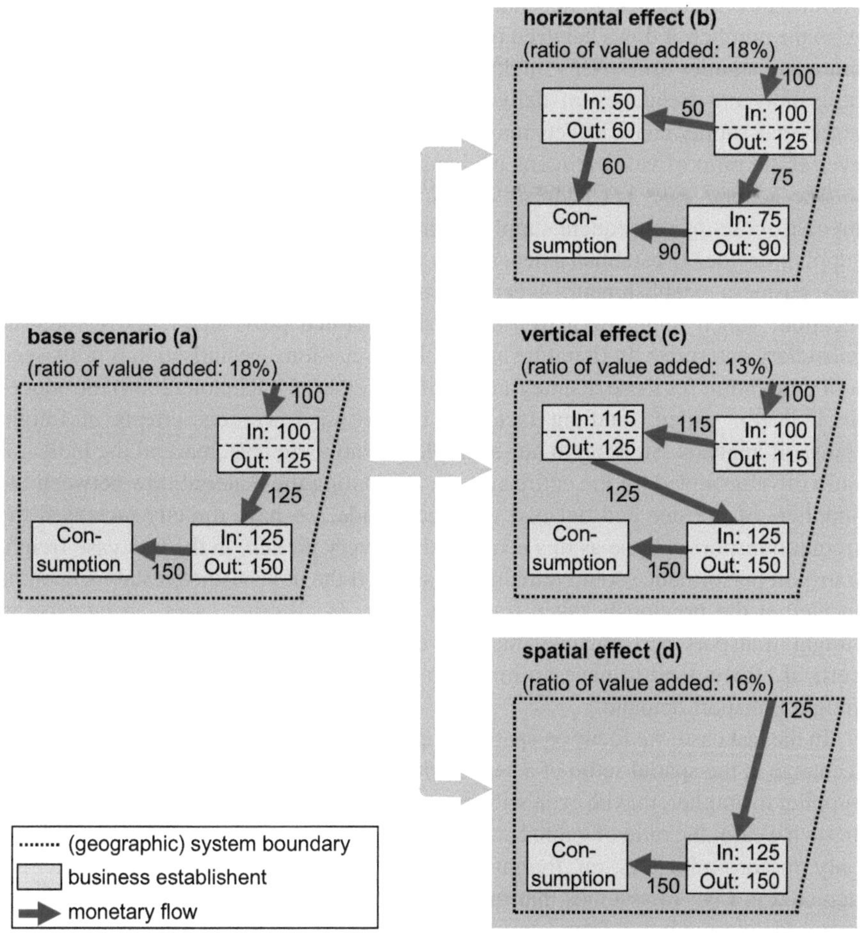

Fig. 4 Examples for supply-chain-structure effects

calculate the overall ratio of value added per output by summing up all procured inputs and outputs and dividing the balance by the total output. Here, it is important to note that the ratio of value added directly depends on the scope of the units to be accounted.

Our first case (b) covers horizontal effects. These are those effects that address actors of the same supply chain stage. In our example the first-tier production is split up and the previously aggregated total volume now is allocated to two business establishments with equal linear production technology. The technology assumption leads to the same amount of input unit required per production of output unit. In reality, similar effects may occur if an OEM decides to switch from single sourcing to double sourcing in order to increase security of supply or to foster competition between suppliers. In the case depicted above, we see no impact on the ratio of valued added for the total system. In contrast, the related material flows have changed. The indicator of freight volume remains equal, but freight transport performance might change depending on the location of the new first-tier supplier. Also the number of flows between business establishments has increased, leading to more and smaller flows. This might influence freight transport aspects like vehicle utilisation and frequency of delivery. For this first case, we can conclude that horizontal supply chain effects may have a significant impact on freight transport even if the ratio of value remains unchanged.

The second case (c) addresses vertical effects. As described, these effects involve actors of consequent supply chain stages. In our example, the second-tier supplier decides to establish a new supply chain stage between tier one and two at a new business establishment. A certain production step is outsourced to that new location, which might be a well-established service provider or newly founded subsidiary enterprise. In literature and public discussion, specialisation is often seen as a motivation for these disintegrations of processes. McKinnon (2008) introduces an indicator called handling factor to capture such vertical effects in freight transport analysis. Sticking to our example, we now see a decrease in the indicator ratio of value added for the entire system. By routing the material flow between the suppliers of tier one and tier two via a new node, we have directly increased the freight transport volume as this is counted on every pickup. In the best case freight transport performance could remain unchanged if the new business establishment is located at the previously taken transport route. In all other cases, an increase in freight transport performance must be expected. Hence we can conclude that vertical effects tend to have an impact on the ratio of value added as well as on freight transport demand.

In the last case, we focus on spatial supply chain effects. A very simple cause for a change in the spatial setup of a supply chain could be a customer's shift from one supplier to another, maybe even within the direct neighbourhood of the old supplier. In such a case, the ratio of valued added remains the same if technology is equal and only minor influences on freight transport demand will occur. The example depicted in Fig. 4d assumes that our first-tier supplier decides to directly procure required intermediates from abroad, expulsing the second-tier supplier location from the supply chain. In such cases in the automotive industry, it is not uncommon that the new offshore plants are run by the same firms as the previously domestic business establishments. Economically speaking, our supply chain now imports more value added (for the economic impact, see OECD 2013) leading to a reduced ratio of value added within our focal system. For freight transport, there can be no doubt that there will be a certain impact, but its manifestation heavily depends on the offshoring location and the previous second-tier supplier's location. Hence, we can conclude that spatial restructurings might be neutral to value added and freight transport if they remain within local boundaries. In case of shifts across international borders value added as well as freight transport will be affected.

Our examples have shown that changes in the ratio of valued added do not uniquely translate into changed freight transport demand. We have also made clear that horizontal, vertical as well as spatial alterations of supply chain structures have effects on freight transport demand. In the following, we will discuss data that can be used in freight transport demand models focusing on the automotive industry that especially address these kinds of effects.

Potential Data Sources for Freight Transport Demand Modelling

During model design it is necessary to decide which parameters need to be endogenous and which can be considered as exogenous. This decision potentially depends on the availability of data as the values of exogenous parameters can either be defined by relying on suitable data or if currently no such data is available assumptions need to be made. In later modelling stages, data from statistics or surveys can also be used for model calibration and verification. Thus, already at an early stage, it is important to get an overview on existing data sources that can be valuable for model creation as well as application.

Since we intend to analyse supply chain structure effects, we require spatial data to assign our system elements, i.e. business establishments, to locations. The car plant locations can easily be taken from publications like Automobil Produktion (2008). This source does not only contain spatial information but also offers additional insight like the number of employees per car plant as well as the type and annual number of cars produced. Getting the locations of the production facilities that supply the car plants is more challenging. Commercial data providers like Dun and Bradstreet (D&B) do have databases on business establishments together with a categorisation by economic sectors. A severe disadvantage when using this data for our purpose is its proprietary character meaning that it is not available publicly and that it is only partly consistent with standards and classifications defined by the official statistical offices. Besides these commercial data providers, there are also directories about automotive suppliers published by interest groups like member lists of the German Association of the Automotive Industry (VDA) or product-related reference books (Automobil Produktion 2011). Due to their origin, they do only cover a random range of the existing companies. In contrast to the previously described private collections of business establishments, full spatial and sectoral coverage is offered by the employment statistics available from the Federal Employment Agency. The advantage of this data's completeness faces the disadvantage of aggregation. Instead of information on the level of single business establishments, we get numbers of employees per economic sector and regional zones. Apart from aggregation, this data suffers from data protection as certain rules are enforced that lead to the blackening of dataset entries in case conclusions on single business establishments could be drawn. Despite all these drawbacks, the statistical employment data allows for the regional distribution of economic activity, manufacturing of goods being a part of this, in Germany down to the level of NUTS-3 regions (corresponding to a diameter of approximately 50 km) and three-digit sector codes. So this is a viable source for the spatial description of freight generation as information on the amount of goods produced can be found in the national production statistics (Eurostat 2013) on a very detailed level of product categories.

Besides the location, we also need information on the size of business establishments. It happens that just this data is available from the German Statistical Office as well, giving the distribution of employees per business establishment per economic sector up to a three-digit classification of economic activity (Statistisches Bundesamt 2012). By combining employment data and size distribution, we can produce an artificial population of business establishments as basic elements of our modelling concept. The next question now must be which data we have to create relations, especially material flows, between the business establishments. On an aggregated level, the national input–output table (Statistisches Bundesamt 2014) gives data on the procurement of intermediates and final consumption for the different economic sectors. One of the many sources used for the creation of the input–output table are input surveys conducted regularly. They offer slightly more detailed information making them helpful for our purpose and are also available publicly (Statistisches Bundesamt 2009). These two sources still leave a blind spot on material flows across borders, which the official foreign trade statistics fill out extensively. From trade statistics we can also derive the value density for the goods traded and transported.

More detailed analyses of business establishment behaviour and characteristics are available in publications on case studies or surveys. For example for the automotive industry, Knorst (2004) describes the relation between an exemplary car plant and its suppliers, Schulz and Hesse (2012) focus on a specific firm's production network and Iddink (2011) addresses the number of suppliers per business establishment and the supplier's spatial distribution based on categories of distance.

First Evidence for Supply Chain Structure Effects

We have also taken a first look on the available data in order to find first evidence for the existence of the supply chain structure effects described previously. Table 1 shows the number of business establishments categorised by the number of employees and supply chain stage for the year 1995 and 2012. Here, the classification NACE Rev. 2 Group 29.1 represents manufacturers' car plants and we assume that Group 29.3 can be directly translated into suppliers from or close to tier one. Now, looking at our data with this in mind, we find the largest growth in the number of small to mid-size business establishments, the strongest increase coming from suppliers in the category of 100–499 employees.

In our interpretation, this growth supports the existence of vertical supply chain effects. Potential reasons for the integration of additional business establishments into the supply chain are conventional outsourcing, modularisation of the assembly process or the increased procurement of complex intermediates in order to keep innovation leadership and integrate technology developed by suppliers.

From the national accounting data for the years 2000 and 2010 for Germany, we get a similar image (see Table 2). At first, we see an overall increase in the total production value of the German automotive industry. Obviously, this is accompanied by an increase in the value of intermediates procured. As a part thereof, we can identify the value of automotive parts bought as intermediates. Supporting our hypotheses of vertical effects, we see that the share of the automotive parts procured

Table 1 Development of the number of business establishments from 1995 to 2012. Data from Statistisches Bundesamt (1995) and Statistisches Bundesamt (2012)

		Number of Business establishments with employees					
Stage[a]	Year	20–49	50–99	100–499	500–999	≥1000	Total
OEM	1995	18	3	16	9	39	85
OEM	2012	18	12	22	16	41	109
Supplier	1995	110	98	233	49	59	549
Supplier	2012	189	173	363	86	56	867
OEM	Growth	±0	+9	+6	+7	+2	+24
Supplier	Growth	+79	+75	+130	+37	−3	+318

[a]Corresponds to NACE Rev. 2 Group 29.1 for OEM and 29.3 for Supplier

Table 2 National accounting data on the German automotive industry for the years 2000 and 2010 (Statistisches Bundesamt 2014)

	2000 (m EUR)	2010 (m EUR)	Growth (%)
Production value	211,897	285,180	35
Inputs total (domestic)	147,664	169,635	15
From automotive industry (domestic)	70,609	86,210	22
Inputs total (import)	26,416	55,200	109
From automotive industry (import)	6720	26,265	291

in relation to all intermediates has significantly grown from 2000 to 2010. This effect holds true for domestic parts (from 48 to 58 %), but it is especially strong for imports (from 25 to 48 %). Comparing domestic and international procurement, we see that imports have grown much stronger than domestic intermediates. This is a hint that supply chains have been subject to spatial effects during the time considered.

Now, we have described potential input data for the intended freight transport demand model and given first insights. So far, this insight is on a very aggregate level and we only have a vague idea how the change of certain aggregate indicators can be explained on supply chain level. Hence, we will present a model concept that can be used to trace supply chain structure effects and their impact on freight transport demand.

Capturing Supply Chain Structure Effects in Freight Transport Demand Models

Originally, freight transport models were built according to passenger transport models following a four-step design as shown in Fig. 5a. In the first step, these models generate incoming and outgoing freight volumes per zone based on structural data. In the second step, gravity models are used in order to calculate interzonal freight flows. Next, for these freight flows transport modes are chosen and vehicle flows are derived. In the fourth step, these vehicle flows get assigned to the transport infrastructure.

Usually, these aggregate four-step models do not include causal relations explaining system elements' behaviour. Therefore, their suitability for the analysis of system behaviour as a reaction towards internal or external influences is limited. A schematic approach how to explanatorily describe freight transport demand as

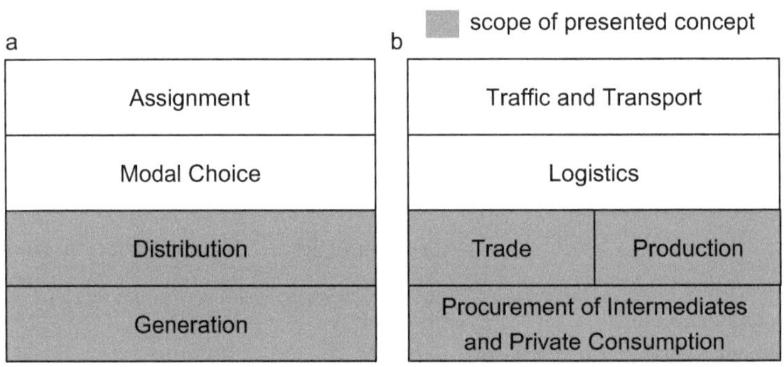

Fig. 5 Traditional four-step approach originating from passenger transport modelling (**a**) and systems determining freight transport demand (**b**)

emerging from behaviour and interaction of subsystems is shown in Fig. 5b building upon Sjöstedt (2004).

As mentioned before, we propose to limit the model scope to a specific sector in order to better capture the relevant drivers of freight transport demand. This is supported by data availability on the sectoral level and the higher homogeneity of behaviour within production stages and economic stakeholders within a branch. Thus, on the sectoral level explanatory modelling becomes possible.

For the automotive industry, we have identified supply chain structure effects as being major drivers for freight transport demand. Therefore, the presented concept focuses on the fundamentals for capturing these effects in a model, which is the spatial distribution of business establishments and the material flows between them. In the terms of four-step modelling, this corresponds to the first two steps, freight generation and freight distribution.

As described, our model needs to be based on business establishments as smallest economic units. The result of our model shall be material flows between these business establishments. Hence, we need an algorithm to derive origin, destination and volumes for the incoming and outgoing material flows for all business establishments based on available input data. In the first step, we can generate business establishments with information on the industry they belong to (e.g. ISIC classification, equal to type of goods produced, e.g. CPA classification), the number of employees they have and the location they are at. The artificial industry landscape, generated by Monte Carlo simulation, is based on company size distributions and disaggregated employment data, which contains information on the number of employees per NUTS-3-region and three-digit NACE classification of economic activities. Similar approaches can be found in Liedtke (2006) and Friedrich (2010). For business establishment representing car plants, precise data on the location, employment and even output are available that can be fed directly into the model bypassing the artificial generation procedure. The output of the first step in the model is a list of all business establishments including their location, number of employees and economic activity (*manufacturing of vehicles or parts thereof* as well as connected supplying industries).

In the next step, we calculate the business establishments' material output. For the automotive supply chains, overall demand is determined by the final output of vehicles from the car plants. Since precise data on the number of vehicles produced per car plant and vehicle type is available, we can disaggregate these vehicles into required modules and components following simplified bills of materials and use this information for the car plants and partly for their suppliers. For the remaining business establishments, the total national production of goods is used to calculate their annual output assuming linear employment productivity. In later versions, this could be improved by including economies of scale derived from already available statistics on average turnover per size category.

Generally, we consider each business establishment as a node that has equilibrium in the masses of input and output flows ($\sum m_{\text{input}} = \sum m_{\text{output}} + \sum m_{\text{material loss}}$). Hence, we can derive the volume of the aggregate input flow based on the output flow, assuming a certain material loss. As business establishments do not get only one but many types of supplies from different sources, we need to break down the aggregate input stream into smaller discrete material flows. We propose to do this breakdown using Monte Carlo simulation, based on expert parameter estimations and literature. In a second step, we assign every material flow an originating industry. Thus, we have a set of material flows, each containing information on the consuming business establishment, the volume and the originating industry. Hence, we can apply a rule-based sourcing for every business establishment and in that way determine the supplying business establishments per material flow. For each type of intermediate certain sourcing rules can be defined following available case or from statistics. The output of this final model step is an artificial industry landscape that contains disaggregate material flows between business establishments. The sourcing algorithm to be applied takes into account data available from statistics and literature as well as industry-specific knowledge. This includes spatial aspects due to reliability and quality aspects as well as strategic constraints like single or dual sourcing.

The model can be used to analyse the impact of supply chain structure effects by applying comparative simulation runs with varying input data and parameter assumptions.

Conclusion and Outlook

In this work, we have pointed out that major parameters of freight transport are set by firms and their customers. It is their decisions that determine which goods have to be transported in which amount and in the consequence which spatial gaps have to be bridged. Often, these decisions are interdependent on logistics, production and transport. Therefore, we consider decisions determining freight transport as being of interdisciplinary nature.

We have also explained why different modelling objectives in these disciplines, combined with a high overall complexity, make it difficult to create a monolithic model. Hence, for freight transport, we have presented the concept of sectoral freight transport demand modelling that allows focusing on the most relevant drivers for the analysis of freight transport within a selected industry. This concept allows for the creation of behaviour-sensitive models, the reduction of complexity and the usage of more detailed data, which is often available on the sector level.

For the German automotive industry, we have shown its prospects for ongoing change as it keeps facing new challenges today and in the future. For this specific industry, we assume that supply chain structure effects are the most significant drivers of freight transport demand. We have categorised these drivers into horizontal, vertical and spatial effects, and by relying on numerical examples, we have shown how these effects influence the ratio of value-added as well as freight transport demand.

In order to capture these effects in freight transport models, we have presented a concept that corresponds to the steps of freight generation and distribution in traditional four-step models. Our approach emphasises on the material flows between business establishments. Hence, we have presented a schematic layout for the procedural generation of such an artificial industry, which can be used for the analysis of supply chain structure effects. Consequently, our next step will be the implementation of the proposed model concept.

For the improvement of freight transport analysis in the long run, our approach can be integrated into existing as well as future full-scale freight transport demand models.

References

Audi. (2008). 15.000 Ganzzüge zwischen den Audi Standorten Győr und Ingolstadt. Accessed May 29, 2012, from http://a8.audi.hu/deu/hirek/hir.php?hir_id=132

Audi. (2013). The production site Győr at a glance. Accessed May 29, 2012, from http://www.audi.com/com/brand/en/company/production_plants/gyoer.html

Automobil Produktion. (2008). *Sonderausgabe Internationale OEM Produktionsstandorte*. Landsberg: mi, verlag moderne industrie.

Automobil Produktion. (2011). *Automobil-Zulieferer in Deutschland 2011/2012*. Landsberg: mi, verlag moderne industrie.

Becker, T. (2006). *Quo Vadis Zulieferpark?: Neue Zusammenarbeitsmodelle in der automobilen Zulieferkette*. Hamburg: Verlag Dr. Kovač.

Bossel, H. (2004). *Systeme, Dynamik, Simulation: Modellbildung, Analyse und Simulation komplexer Systeme*. Norderstedt: Books on Demand.

Bundesministerium für Verkehr, Bau und Stadtentwicklung, (Ed.). (2013). Verkehr in Zahlen (42nd edn). Hamburg: Deutscher Verkehrs-Verlag.

Chow, J. Y., Yang, C. H., & Regan, A. C. (2010). State-of-the-art of freight forecast modeling: Lessons learned and the road ahead. *Transportation: Planning, Policy, Research, Practice, 37*(6), 1011–1030.

Christopher, M. (1992). *Logistics and supply chain management: Strategies for reducing costs and improving services*. London: Financial Times/Pitman Pub.

de Dios Ortúzar, J., & Willumsen, L. G. (2011). *Modelling transport*. Oxford: Wiley-Blackwell.

de Jong, G., & Ben-Akiva, M. (2007). A micro-simulation model of shipment size and transport chain choice. *Transportation Research Part B: Methodological, 41*(9), 950–965.

de Jong, G., Vierth, I., Tavasszy, L., & Ben-Akiva, M. (2013). Recent developments in national and international freight transport models within Europe. *Transportation, 40*(2), 347–371.

Dyer, J. H. (1996). Specialized supplier networks as a source of competitive advantage: Evidence from the auto industry. *Strategic Management Journal, 17*(4), 271–291.

European Commission. (2011). *WHITE PAPER*. Roadmap to a Single European Transport Area: Towards a competitive and resource efficient transport system. Brussels. Accessed August 6, 2011, from http://eur-lex.europa.eu/legal-content/EN/TXT/PDF/?uri=CELEX:52011DC0144&from=EN

Eurostat. (2013). Production statistics (Prodcom). Accessed February 27, 2013, from http://epp.eurostat.ec.europa.eu/portal/page/portal/prodcom/introduction

Friedrich, H. (2010). *Simulation of logistics in food retailing for freight transportation analysis*. Dissertation, Karlsruhe Institute of Technology, Karlsruhe

Göpfert, I., & Grünert, M. (2012). Logistiknetze der Zukunft: Das neue Hersteller-Zulieferer-Verhältnis in der Automobilindustrie. In I. Göpfert (Ed.), *Logistik der Zukunft: Logistics for the future* (pp. 139–178). Wiesbaden: Gabler.

Güttner, A., & Sommer-Dittrich, T. (2008). Current issues at OEMs and suppliers. In G. Parry & A. Graves (Eds.), *Build to order: The road to the 5-day car* (pp. 55–73). London: Springer.

Holweg, M., & Pil, F. K. (2004). *The second century: Reconnecting customer and value chain through build-to-order: Moving beyond mass and lean production in the auto industry*. Cambridge, MA: MIT Press.

Holzapfel, H., & Vahrenkamp, R. (1999). An appraisal of decreased depth of production on traffic demand: Development of a model. *World Transport Policy and Practice, 5*(2), 18–20.

Hutchinson, B. (1974). *Principles of urban transport systems planning*. Washington, DC: Scripta Book Co.

Iddink, U. (2011). Improving freight transport models by integration of logistics and procurement figures. In Transportation Research Board (Ed.), *Transportation Research Board 90th Annual Meeting*. Washington, DC: Transportation Research Board.

Ihme, J. (2006). *Logistik im Automobilbau: Logistikkomponenten und Logistiksysteme im Fahrzeugbau*. München: Hanser.

Katayama, H., & Bennett, D. (1999). Agility, adaptability and leanness: A comparison of concepts and a study of practice. *International Journal of Production Economics, 60–61*(1), 43–51.

KBA. (2013). Neuzulassungen von Personenkraftwagen im Jahr {2008, 2009, 2010} nach Bundesländern und Kraftstoffarten absolut. Flensburg. Accessed February 27, 2013, from http://www.kba.de

Knorst, P. (2004). Ford: Lean manufacturing and supply. In D. Corsten & C. Gabriel (Eds.), *Supply chain management erfolgreich umsetzen: Grundlagen, Realisierung und Fallstudien* (pp. 73–86). Berlin: Springer.

Law, A. (2006). *Simulation modeling and analysis*. New York: McGraw-Hill Publishing Co.

Liedtke, G. (2006). An actor-based approach to commodity transport modelling. Dissertation, Nomo, Baden-Baden.

Liedtke, G., & Friedrich, H. (2012). Generation of logistics networks in freight transportation models. *Transportation, 39*(6), 1335–1351.

MacCarthy, B. L., & Atthirawong, W. (2003). Factors affecting location decisions in international operations—A Delphi study. *International Journal of Operations and Production Management, 23*(7), 794–818.

Manheim, M. L. (1980). *Fundamentals of transportation systems analysis*. Cambridge, MA: MIT Press.

Martin, B. V., Memmott, F. W., & Bone, A. J. (1965). *Principles and techniques of predicting future demand for urban area transportation*. Cambridge, MA: MIT Press.

McKinnon, A. C. (2008). The potential of economic incentives to reduce CO_2 emissions from goods transport. In OECD (Ed.), *International transport forum on transport and energy, Leipzig, 28.5.2008* (pp. 28–30).

OECD. (2013). Measuring trade in value-added: OECD-WTO database on trade in value-added. http://www.oecd.org/sti/ind/TIVA%20flyer%20FINAL.pdf

Page, B., & Kreutzer, W. (2005). *The Java simulation handbook: Simulating discrete event systems with UML and Java (Berichte Aus Der Informatik)*. Aachen: Shaker Verlag GmbH.

Porter, M. E. (1985). *Competitive advantage: Creating and sustaining superior performance*. New York: Collier Macmillan.

PwC. (2012). Europas Pkw-Markt im Rückwärtsgang—Absatz dürfte 2012 um fünf Prozent sinken. Accessed February 27, 2013, from http://www.pwc.de/de/pressemitteilungen/2012/europas-pkw-markt-im-rueckwaertsgang.jhtml

Rogers, G. G., & Bottaci, L. (1997). Modular production systems: A new manufacturing paradigm. *Journal of Intelligent Manufacturing, 8*(2), 147–156.

Roorda, M. J., Cavalcante, R., McCabe, S., & Kwan, H. (2010). A conceptual framework for agent-based modelling of logistics services. *Transportation Research Part E: Logistics and Transportation Review, 46*(1), 18–31.

Samimi, A., Mohammadian, A. K., & Kawamura, K. (2010). Freight demand microsimulation in the U.S. In World Conference on Transport Research Society (Ed.), *General Proceedings of the 12th World Conference on Transport Research Society*. Lisbon.

Schoemaker, P. J. H. (1993). Multiple scenario development: Its conceptual and behavioral foundation. *Strategic Management Journal, 14*(3), 193–213.

Schulz, R., & Hesse, F. (2012). Das Produktionsnetzwerk des VW-Konzerns und die Versorgung der Überseewerke. In I. Göpfert (Ed.), *Logistik der Zukunft: Logistics for the future* (pp. 213–234). Wiesbaden: Gabler.

Simon, H. A. (1996). *The sciences of the artificial*. Cambridge, MA: MIT Press.

Sjöstedt, L. (2004). A conceptual framework for analyzing policy-maker's and industry roles. In P. Rietveld (Ed.), *Barriers to sustainable transport*. London: Taylor and Francis Group.

Statistisches Bundesamt. (1995). Betriebe, Tätige Personen und Umsatz des Verarbeitenden Gewerbes sowie des Bergbaus und der Gewinnung von Steinen und Erden nach Beschäftigtengrößenklassen (Fachserie 4, Reihe 4.1.2). Wiesbaden.

Statistisches Bundesamt. (2009). Material- und Wareneingangserhebung im Verarbeitenden Gewerbe sowie im Bergbau und in der Gewinnung von Steinen und Erden (Fachserie 4 Reihe 4.2.4). Wiesbaden.

Statistisches Bundesamt. (2012). Betriebe, Tätige Personen und Umsatz des Verarbeitenden Gewerbes sowie des Bergbaus und der Gewinnung von Steinen und Erden nach Beschäftigtengrößenklassen (Fachserie 4, Reihe 4.1.2). Wiesbaden.

Statistisches Bundesamt. (2013a). Beförderungsleistung im Inland nach Verkehrsträgern und Güterabteilungen 2011 (NST 2007). Wiesbaden. Accessed February 27, 2013, from https://www.destatis.de/DE/ZahlenFakten/Wirtschaftsbereiche/TransportVerkehr/Gueterverkehr/Tabellen/VerkehrstraegerGueterabteilungB.html

Statistisches Bundesamt. (2013b). Verkehr aktuell 01/2013. Wiesbaden. Accessed February 27, 2013, from https://www.destatis.de/DE/Publikationen/Thematisch/TransportVerkehr/Querschnitt/VerkehrAktuellPDF_2080110.pdf

Statistisches Bundesamt. (2014). Volkswirtschaftliche Gesamtrechnungen (Fachserie 18 Reihe 2). Wiesbaden.

Taniguchi, E., Thompson, R. G., & Yamada, T. (2014). Recent trends and innovations in modelling city logistics. *Procedia—Social and Behavioral Sciences, 125*(1), 4–14.

Tavasszy, L., & Ruijgrok, K. (2013). Freight transport: Indicators, determinants and drivers of change. In B. van Wee, J. A. Annema, & D. Banister (Eds.), *The transport system and transport policy: An introduction* (pp. 51–77). Cheltenham: Edward Elgar Publishing.

Tavasszy, L. A., Ruijrok, K., & Davydenko, I. (2012). Incorporating logistics in freight transport demand models: State-of-the-art and research opportunities. *Transport Reviews, 32*(2), 203–219.

Tillmann, S. (2009). Autoindustrie trickst bei Jobzahlen. *Capital* (06/2009).

VDA. (1969–2012). *Tatsachen und Zahlen*. Berlin: Verband der Automobil industrie.

VDA. (2004). *Die deutsche Automobilindustrie in der erweiterten EU: Motor der Integration*. Berlin: Verband der Automobilindustrie.

Werner, H. (2013). *Supply chain management: Grundlagen, Strategien, Instrumente und Controlling*. Wiesbaden: Springer Gabler.

MIX
Papier aus verantwortungsvollen Quellen
Paper from responsible sources
FSC® C105338

If you have any concerns about our products,
you can contact us on
ProductSafety@springernature.com

In case Publisher is established outside the EU,
the EU authorized representative is:
Springer Nature Customer Service Center GmbH
Europaplatz 3, 69115 Heidelberg, Germany

Printed by Libri Plureos GmbH
in Hamburg, Germany